ELEMENTARY NUMBER THEORY, GROUP THEORY, AND RAMANUJAN GRAPHS

LONDON MATHEMATICAL SOCIETY STUDENT TEXTS

Managing editor: Professor C.M. Series, Mathematics Institute
University of Warwick, Coventry CV4 7AL, United Kingdom

ELEMENTARY NUMBER THEORY, GROUP THEORY, AND RAMANUJAN GRAPHS

GIULIANA DAVIDOFF
Mount Holyoke College

PETER SARNAK
Princeton University & NYU

ALAIN VALETTE
Universite de Neuchatel

CAMBRIDGE
UNIVERSITY PRESS

CAMBRIDGE UNIVERSITY PRESS
Cambridge, New York, Melbourne, Madrid, Cape Town, Singapore,
São Paulo, Delhi, Dubai, Tokyo

Cambridge University Press
The Edinburgh Building, Cambridge CB2 8RU, UK

Published in the United States of America by Cambridge University Press, New York

www.cambridge.org
Information on this title: www.cambridge.org/9780521531436

First published 2003

A catalogue record for this publication is available from the British Library

Library of Congress Cataloguing in Publication data
Davidoff, Giuliana P.
Elementary number theory, group theory, and Ramanujan graphs / Giuliana Davidoff,
Peter Sarnak, Alain Valette.
p.   cm. – (London Mathematical Society student texts; 55)
Includes bibliographical references and index.
ISBN 0-521-82426-5 – ISBN 0-521-53143-8 (pb.)
1. Graph theory.   2. Number theory.   3. Group theory.   I. Sarnak, Peter.
II. Valette, Alain.   III. Title.   IV. Series.
QA166 .D35   2003
511′.5 – dc21   2002074057

ISBN 978-0-521-82426-2 Hardback
ISBN 978-0-521-53143-6 Paperback

Transferred to digital printing 2010

Contents

Preface

These notes are intended for a general mathematical audience. In particular, we have in mind that they could be used as a course for undergraduates. They contain an explicit construction of highly connected but sparse graphs known as expander graphs. Besides their interest in combinatorics and graph theory, these graphs have applications to computer science and engineering. Our aim has been to give a self-contained treatment. Thus, the relevant background material in graph theory, number theory, group theory, and representation theory is presented. The text can be used as a brief introduction to these modern subjects as well as an example of how such topics are synthesized in modern mathematics. Prerequisites include linear algebra together with elementary algebra, analysis, and combinatorics.

Giuliana Davidoff
Department of Mathematics
Mount Holyoke College
South Hadley, MA

An Overview

In this book, we shall consider graphs $X = (V, E)$, where V is the set of vertices and E is the set of edges of X. We shall assume that X is undirected; most of the time, X will be finite. A *path* in X is a sequence $v_1 v_2 \ldots v_k$ of vertices, where v_i is adjacent to v_{i+1} (i.e., $\{v_i, v_{i+1}\}$ is an edge). A graph X is *connected* if every two vertices can be joined by a path.

For $F \subseteq V$, the *boundary* ∂F is the set of edges connecting F to $V - F$. Consider for example the graph in Figure 0.1 (this is the celebrated Petersen graph): it has 10 vertices and 15 edges; three vertices have been surrounded by squares: this is our subset F; the seven "fat" edges are the ones in ∂F.

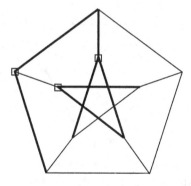

Figure 0.1

The *expanding constant*, or *isoperimetric constant* of X, is

$$h(X) = \inf \left\{ \frac{|\partial F|}{\min\{|F|, |V - F|\}} : F \subseteq V : 0 < |F| < +\infty \right\}.$$

If we view X as a network transmitting information (where information retained by some vertex propagates, say in 1 unit of time, to neighboring vertices), then $h(X)$ measures the "quality" of X as a network: if $h(X)$ is large,

1

information propagates well. Let us consider two extreme examples to illustrate this.

0.1.1. Example. The *complete graph* K_m on m vertices is defined by requiring every vertex to be connected to any other, distinct vertex: see Figure 0.2 for $m = 5$.

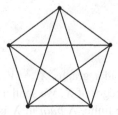

Figure 0.2

It is clear that, if $|F| = \ell$, then $|\partial F| = \ell(m - \ell)$, so that $h(K_m) = m - \left[\frac{m}{2}\right] \sim \frac{m}{2}$.

0.2.2. Example. The *cycle C_n* on n vertices: see Figure 0.3 for $n = 6$. If F is a half-cycle, then $|\partial F| = 2$, so $h(C_n) \leq \frac{2}{\left[\frac{n}{2}\right]} \sim \frac{4}{n}$; in particular $h(C_n) \to 0$ for $n \to +\infty$.

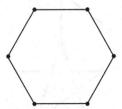

Figure 0.3

From these two examples, wee see that the highly connected complete graph has a large expanding constant that grows proportionately with the number of vertices. On the other hand, the minimally connected cycle graph has a small expanding constant that decreases to zero as the number of vertices grows. In this sense, $h(X)$ does indeed provide a measure of the "quality," or connectivity of X as a network.

We say that a graph X is *k-regular* if every vertex has exactly k neighbors, so that the Petersen graph is 3-regular, K_m is $(m - 1)$-regular, and C_n is 2-regular.

0.3.3. Definition. Let $(X_m)_{m \geq 1}$ be a family of graphs $X_m = (V_m, E_m)$ indexed by $m \in \mathbb{N}$. Furthermore, fix $k \geq 2$. Such a family $(X_m)_{m \geq 1}$ of finite, connected,

k-regular graphs is a *family of expanders* if $|V_m| \to +\infty$ for $m \to +\infty$, and if there exists $\varepsilon > 0$, such that $h(X_m) \geq \varepsilon$ for every $m \geq 1$.

Because an optimal design for a network should take economy of transmission into account, we include the assumption that X_m is k-regular in Definition 0.3.3. This assures that the number of edges of X_m grows linearly with the number of vertices. Without that assumption, we could just take $X_m = K_m$ for good connectivity. However, note that K_m has $\frac{m(m-1)}{2}$ edges, which quickly becomes expensive when transmission lines are made of either copper or optical fibers. Hence, the "optimal" network for practical purposes arises from a graph that provides the best connectivity from a minimal number of edges.

Indeed such expander graphs have become basic building blocks in many engineering applications. We cite a few such applications, taken from Reingold, Vadhan and Wigderson [55]: to network designs [53], to complexity theory [66], to derandomization [50], to coding theory [63], and to cryptography [30].

0.4.4. Main Problem. Give explicit constructions for families of expanders.

We shall solve this problem algebraically, by appealing to the *adjacency matrix* A of the graph $X = (V, E)$; it is indexed by pairs of vertices x, y of X, and A_{xy} is the number of edges between x and y.

When X has n vertices, A is an n-by-n, symmetric matrix, which completely determines X. By standard linear algebra, A has n real eigenvalues, repeated according to multiplicities that we list in decreasing order:

$$\mu_0 \geq \mu_1 \geq \cdots \geq \mu_{n-1}.$$

In section 1.1 we shall prove the following.

0.5.5. Proposition. If X is a k-regular graph on n vertices, then

$$\mu_0 = k \geq \mu_1 \geq \cdots \geq \mu_{n-1} \geq -k.$$

Moreover,

(a) $\mu_0 > \mu_1$ if and only if X is connected.
(b) Suppose X is connected. The equality $\mu_{n-1} = -k$ holds if and only if X is bicolorable. (A graph X is *bicolorable* if it is possible to paint the vertices of X in two colors in such a way that adjacent vertices have distinct colors.)

It turns out that the expanding constant can be estimated spectrally by means of a double inequality (due to Alon & Milman [3] and to Dodziuk [22]) that we shall prove in section 1.2.

0.6.6. Theorem. Let X be a finite, connected, k-regular graph. Then

$$\frac{k - \mu_1}{2} \leq h(X) \leq \sqrt{2k\,(k - \mu_1)}\,.$$

This allows for an equivalent formulation of 0.4.4.

0.7.7. Rephrasing of the Main Problem. Give explicit constructions for families $(X_m)_{m \geq 1}$ of finite, connected, k-regular graphs with the following properties: (i) $|V_m| \to +\infty$ for $m \to +\infty$, and (ii) there exists $\varepsilon > 0$ such that $k - \mu_1(X_m) \geq \varepsilon$ for every $m \geq 1$.

Therefore, to have good quality expanders, the *spectral gap* $k - \mu_1(X_m)$ has to be as large as possible. However, the spectral gap cannot be too large as was observed independently by Alon and Boppana [10] and Serre [62] (see also Grigorchuk & Zuk [31]). In fact, we have the bound implied by the following result.

0.8.8. Theorem. Let $(X_m)_{m \geq 1}$ be a family of finite, connected, k-regular graphs with $|V_m| \to +\infty$ as $m \to +\infty$. Then

$$\liminf_{m \to +\infty} \mu_1(X_m) \geq 2\sqrt{k - 1}\,.$$

This asymptotic threshold will be discussed in section 1.3 and proved in section 1.4. Now Theorem 0.8.8 singles out an extremal property on the eigenvalues of the adjacency matrix of a k-regular graph; this motivates the definition of a Ramanujan graph.

0.9.9. Definition. A finite, connected, k-regular graph X is *Ramanujan* if, for every eigenvalue μ of A other than $\pm k$, one has

$$|\mu| \leq 2\sqrt{k - 1}\,.$$

So, if for some $k \geq 3$ we succeed in constructing an infinite family of k-regular Ramanujan graphs, we will get a solution of our main problem 0.7.7 (hence, also of 0.4) which is optimal from the spectral point of view.

0.10.10. Theorem. For the following values of k, there exist infinite families of k-regular Ramanujan graphs:

- $k = p + 1$, where p is an odd prime ([42], [46]).
- $k = 3$ [14].
- $k = q + 1$, where q is a prime power [48].

Our purpose in this book is to describe the Ramanujan graphs of Lubotzky et al. [42] and Margulis [46]. While the description of these Ramanujan graphs (given in section 4.2) is elementary, the proof that they have the desired properties is not. For example, the proofs in [42] and [41] make free use of the theory of algebraic groups, modular forms, theta correspondences, and the Riemann Hypothesis for curves over finite fields. Our aim here is to give elementary and self-contained proofs of most of the properties enjoyed by these graphs, results the reader will find in sections 4.3 and 4.4. Actually, our elementary methods will not give us the full strength of the Ramanujan bound for these graphs, though they do have that property. Nevertheless, we will be able to prove that they form a family of expanders with a quite good explicit asymptotic estimate on the spectral gap. This estimate is strong enough to provide explicit solutions to two outstanding problems in graph theory that we describe as follows:

0.11.11. Definition. Let X be a graph.

(a) The *girth* of X, denoted by $g(X)$, is the length of the shortest circuit in X.

(b) The *chromatic number* of X, denoted by $\chi(X)$, is the minimal number of colors needed to paint the vertices of X in such a way that adjacent vertices have different colors.

The problem of the existence of finite graphs with large girth and at the same time large chromatic number has a long history (see [7]). The problem was first solved by Erdös [24], whose solution shows that the "random graph" has this property; this construction is recalled in section 1.7. (This paper was the genesis of the "random method" and theory of random graphs. See the monograph [4].) We shall see in section 4.4 that the graphs $X^{p,q}$ presented in Chapter 4 provide explicit solutions to this problem.

0.12.12. Definition. Let $(X_m)_{m \geq 1}$ be a family of finite, connected, k-regular graphs, with $|V_m| \to +\infty$ as $m \to +\infty$. We say that this family has *large girth* if, for some constant $C > 0$, one has $g(X_m) \geq (C + o(1)) \log_{k-1} |V_m|$, where $o(1)$ is a quantity tending to 0 for $m \to +\infty$.

It is easy to see that, necessarily, $C \leq 2$. By counting arguments, Erdös and Sachs [25] proved the existence of families of graphs with large girth and with $C = 1$. In the Appendix, we give a beautiful explicit construction due to Margulis [45], leading to $C = \frac{1}{3} \frac{\log 3}{\log(1+\sqrt{2})} = 0.415 \ldots$. In section 4.3, we shall see that the graphs $X^{p,q}$, with p not a square modulo q, provide a family with large girth and that $C = \frac{4}{3}$ which, asymptotically, is the largest girth known.

We claimed previously that our constructions are "elementary": since there is no general agreement on the meaning of this word, we feel committed to clarify it somewhat. In 1993, the first two authors wrote up a set of unpublished Notes that were circulated under the title "An elementary approach to Ramanujan graphs." In 1998–99, the third author based an undergraduate course on these Notes; in the process he was able to simplify the presentation even further. This gave the impetus for the present text. We assume that our reader is familiar with linear algebra, congruences, finite fields of prime order, and some basic ring theory. The relevant number theory is presented in Chapter 2; and the group theory, including representation theory, in Chapter 3.

Other than these topics, we have attempted to present here a self-contained treatment of the construction and proofs involved. To do this we have borrowed some of our exposition from well-known sources, adapting and tailoring those to give a more concise presentation of the contexts and specific theoretical tools we need. In all such cases, we hope that we have provided clear and complete attribution of sources for those readers who wish to pursue any topic more broadly.

There is some novelty in our approach.

- The graphs $X^{p,q}$ depend on two distinct, odd primes p, q. In the literature, it is commonly assumed that $p \equiv 1 \pmod{4}$, for simplicity. We give a complete treatment of both the case $p \equiv 1 \pmod{4}$ and the case $p \equiv 3 \pmod{4}$.

- As in [42], [44], and [57], we give two constructions of the graphs $X^{p,q}$: one is based on quaternion algebras and produces connected graphs by construction; however, it gives little information about the number of vertices; the other describes the $X^{p,q}$ as Cayley graphs of $\mathrm{PGL}_2(q)$ or $\mathrm{PSL}_2(q)$, from which the number of vertices is obvious but connectedness is not. The isomorphism of both constructions, in the original paper [42] (and also in Proposition 3.4.1 in [57]), depends on fairly deep results of Mališev [43] on the Hardy–Littlewood theory of quadratic forms. The proof in Theorem 7.4.3 of [41] appeals to Kneser's strong approximation theorem for algebraic groups over the adèles. In our approach here, we first give *a priori* estimates on the girth of the graphs obtained by the first method, showing that the girth cannot be too small. We then apply a result of Dickson [20], reproved in section 3.3, that up to two exceptions, proper subgroups of $\mathrm{PSL}_2(q)$ are metabelian, so that Cayley graphs of proper subgroups must have small girth. This is enough to conclude that our Cayley graphs of $PGL_2(q)$ or $PSL_2(q)$ must be connected.

- The proof we give here that the $X^{p,q}$'s, with fixed p, form a family of expanders depends on a result going back to Frobenius [27], and is proved in section 3.5: any nontrivial representation of $PSL_2(q)$ has degree at least $\frac{q-1}{2}$. As a consequence, the multiplicity of any nontrivial eigenvalue of $X^{p,q}$ is at least $\frac{q-1}{2}$. Using the fact that $\frac{q-1}{2}$ is fairly large compared to q^3, the approximate number of vertices, we deduce that there must be a spectral gap.

The idea of trying to exploit this feature of the representations of $PSL_2(q)$ was suggested by Bernstein and Kazhdan (see [8] and [58]). In Sarnak and Xue [59], this lower bound for the multiplicity is combined with some upper-bound counting arguments to rule out exceptional eigenvalues of quotients of the Lobachevski upper half-plane by congruence subgroups in co-compact arithmetic lattices in $SL_2(\mathbb{R})$. Our proof of the spectral gap in these notes is based on similar ideas. This method has also been used recently by Gamburd [29] to establish a spectral gap property for certain families of infinite index subgroups of $SL_2(\mathbb{Z})$.

Most of the exercices in this book were provided by Nicolas Louvet, who was the third author's teaching assistant: we heartily thank him for that. We also thank J. Dodziuk, F. Labourie, F. Ledrappier, and J.-P. Serre for useful comments, conversations, and correspondence.

The draft of this book was completed during a stay of the first author at the University of Roma La Sapienza and of the third author at IHES in the Fall of 1999. It was also at IHES that the book was typed, with remarkable efficiency, by Mrs Cécile Gourgues. We thank her for her beautiful job.

Chapter 1
Graph Theory

1.1. The Adjacency Matrix and Its Spectrum

We shall be concerned with graphs $X = (V, E)$, where V is the set of vertices and E is the set of edges. As stated in the Overview, we always assume our graphs to be undirected, and most often we will deal with finite graphs.

We let $V = \{v_1, v_2, \ldots\}$ be the set of vertices of X. Then the *adjacency matrix* of the graph X is the matrix A indexed by pairs of vertices $v_i, v_j \in V$. That is, $A = (A_{ij})$, where

$$A_{ij} = \text{number of edges joining } v_i \text{ to } v_j.$$

We say that X is *simple* if there is at most one edge joining adjacent vertices; hence, X is simple if and only if $A_{ij} \in \{0, 1\}$ for every $v_i, v_j \in V$.

Note that A completely determines X and that A is symmetric because X is undirected. Furthermore, X has no loops if and only if $A_{ii} = 0$ for every $v_i \in V$.

1.1.1. Definition. Let $k \geq 2$ be an integer. We say that the graph X is *k-regular* if for every $v_i \in V : \sum_{v_j \in V} A_{ij} = k$.

If X has no loop, this amounts to saying that each vertex has exactly k neighbors.

Assume that X is a finite graph on n vertices. Then A is an n-by-n symmetric matrix; hence, it has n real eigenvalues, counting multiplicities, that we may list in decreasing order:

$$\mu_0 \geq \mu_1 \geq \cdots \geq \mu_{n-1}.$$

The *spectrum* of X is the set of eigenvalues of A. Note that μ_0 is a simple eigenvalue, or has multiplicity 1, if and only if $\mu_0 > \mu_1$.

8

For an arbitrary graph $X = (V, E)$, consider functions $f : V \to \mathbb{C}$ from the set of vertices of X to the complex numbers, and define

$$\ell^2(V) = \{f : V \to \mathbb{C} : \sum_{v \in V} |f(v)|^2 < +\infty\}.$$

The space $\ell^2(E)$ is defined analogously.

Clearly, if V is finite, say $|V| = n$, then every function $f : V \to \mathbb{C}$ is in $\ell^2(V)$. We can think of each such function as a vector in $\mathbb{C}^n$ on which the adjacency matrix acts in the usual way:

$$Af = \begin{pmatrix} A_{11} & A_{12} & \cdots & A_{1n} \\ \vdots & \vdots & & \vdots \\ A_{i1} & A_{i2} & \cdots & A_{in} \\ \vdots & \vdots & & \vdots \\ A_{n1} & A_{n2} & \cdots & A_{nn} \end{pmatrix} \begin{pmatrix} f(v_1) \\ f(v_2) \\ \vdots \\ f(v_n) \end{pmatrix}$$

$$= \begin{pmatrix} A_{11} f(v_1) + A_{12} f(v_2) + \cdots + A_{1n} f(v_n) \\ \vdots \\ A_{i1} f(v_1) + A_{i2} f(v_2) + \cdots + A_{in} f(v_n) \\ \vdots \\ A_{n1} f(v_1) + A_{n2} f(v_2) + \cdots + A_{nn} f(v_n) \end{pmatrix}.$$

Hence, $(Af)(v_i) = \sum_{j=1}^{n} A_{ij} f(v_j)$. It is very convenient, both notationally and conceptually, to forget about the numbering of vertices and to index matrix entries of A directly by pairs of vertices. So we shall represent A by a matrix $(A_{xy})_{x,y \in V}$, and the previous formula becomes $(Af)(x) = \sum_{y \in V} A_{xy} f(y)$, for every $x \in V$.

1.1.2. Proposition. Let X be a finite k-regular graph with n vertices. Then

(a) $\mu_0 = k$;
(b) $|\mu_i| \le k$ for $1 \le i \le n - 1$;
(c) μ_0 has multiplicity 1, if and only if X is connected.

Proof. We prove (a) and (b) simultaneously by noticing first that the constant function $f \equiv 1$ on V is an eigenfunction of A associated with the eigenvalue k. Next, we prove that, if μ is any eigenvalue, then $|\mu| \le k$. Indeed, let f be a real-valued eigenfunction associated with μ. Let $x \in V$ be such that

$$|f(x)| = \max_{y \in V} |f(y)|.$$

Replacing f by $-f$ if necessary, we may assume $f(x) > 0$. Then

$$f(x)|\mu| = |f(x)\mu| = \left|\sum_{y \in V} A_{xy} f(y)\right| \leq \sum_{y \in V} A_{xy} |f(y)|$$

$$\leq f(x) \sum_{y \in V} A_{xy} = f(x) k.$$

Cancelling out $f(x)$ gives the result.

To prove (c), assume first that X is connected. Let f be a real-valued eigenfunction associated with the eigenvalue k. We have to prove that f is constant. As before, let $x \in V$ be a vertex such that $|f(x)| = \max_{y \in V} |f(y)|$. As $f(x) = \frac{(Af)(x)}{k} = \sum_{y \in V} \frac{A_{xy}}{k} f(y)$, we see that $f(x)$ is a convex combination of real numbers which are, in modulus, less than $|f(x)|$. This implies that $f(y) = f(x)$ for every $y \in V$, such that $A_{xy} \neq 0$, that is, for every y adjacent to x. Then, by a similar argument, f has the same value $f(x)$ on every vertex adjacent to such a y, and so on. Since X is connected, f must be constant.

We leave the proof of the converse as an exercise. $\square$

Proposition 1.1.2(c) shows a first connection between spectral properties of the adjacency matrix and combinatorial properties of the graph. This is one of the themes of this chapter.

1.1.3. Definition. A graph $X = (V, E)$ is *bipartite*, or *bicolorable*, if there exists a partition of the vertices $V = V_+ \cup V_-$, such that, for any two vertices x, y with $A_{xy} \neq 0$, if $x \in V_+$ (resp. V_-), then $y \in V_-$ (resp. V_+).

In other words, it is possible to paint the vertices with two colors in such a way that no two adjacent vertices have the same color. Bipartite graphs have very nice spectral properties characterized by the following:

1.1.4. Proposition. Let X be a connected, k-regular graph on n vertices. The following are equivalent:

  (i) X is bipartite;
  (ii) the spectrum of X is symmetric about 0;
  (iii) $\mu_{n-1} = -k$.

Proof.

  (i) $\Rightarrow$ (ii) Assume that $V = V_+ \cup V_-$ is a bipartition of X. To show symmetry of the spectrum, we assume that f is an eigenfunction of A with associated eigenvalue μ. Define

$$g(x) = \begin{cases} f(x) & \text{if } x \in V_+ \\ -f(x) & \text{if } x \in V_- \end{cases}.$$

It is then straightforward to show that $(Ag)(x) = -\mu\, g(x)$ for every $x \in V$.

  (ii) $\Rightarrow$ (iii) This is clear from Proposition 1.1.2.

  (iii) $\Rightarrow$ (i) Let f be a real-valued eigenfunction of A with eigenvalue $-k$.

Let $x \in V$ be such that $|f(x)| = \max_{y \in V} |f(y)|$. Replacing f by $-f$ if necessary, we may assume $f(x) > 0$. Now

$$f(x) = -\frac{(Af)(x)}{k} = -\sum_{y \in V} \frac{A_{xy}}{k} f(y) = \sum_{y \in V} \frac{A_{xy}}{k} (-f(y)).$$

So $f(x)$ is a convex combination of the $-f(y)$'s which are, in modulus, less than $|f(x)|$. Therefore, $-f(y) = f(x)$ for every $y \in V$, such that $A_{xy} \neq 0$, that is, for every y adjacent to x. Similarly, if z is a vertex adjacent to any such y, then $f(z) = -f(y) = f(x)$. Define $V_+ = \{y \in V : f(y) > 0\}$, $V_- = \{y \in V : f(y) < 0\}$; because X is connected, this defines a bipartition of X. $\square$

    Thus, every finite, connected, k-regular graph X has largest positive eigenvalue $\mu_0 = k$; if, in addition, X is bipartite, then the eigenvalue $\mu_{n-1} = -k$ also occurs (and only in this case). These eigenvalues k and $-k$, if the second occurs, are called the *trivial* eigenvalues of X. The difference $k - \mu_1 = \mu_0 - \mu_1$ is the *spectral gap* of X.

Exercises on Section 1.1

1. For the complete graph K_n and the cycle C_n, write down the adjacency matrix and compute the spectrum of the graph (with multiplicities). When are these graphs bipartite?

2. Let D_n be the following graph on $2n$ vertices: $V = \mathbb{Z}/n\mathbb{Z} \times \{0, 1\}$; $E = \{\{(i, j), (i + 1, j) : i \in \mathbb{Z}/n\mathbb{Z}, j \in \{0, 1\}\} \cup \{\{(i, 0), (i, 1)\} : i \in \mathbb{Z}/n\mathbb{Z}\}$. Make a drawing and repeat exercise 1 for D_n.

3. Show that a graph is bipartite if and only if it has no circuit with odd length.

4. Let X be a finite, k-regular graph. Complete the proof of Proposition 1.1.2 by showing that the multiplicity of the eigenvalue k is equal to the number of connected components of X (Hint: look at the space of locally constant functions on X.)

5. Let X be a finite, simple graph without loop. Assume that, for some $r \geq 2$, it is possible to find a set of r vertices all having the same neighbors. Show that 0 is an eigenvalue of A, with multiplicity at least $r - 1$.

6. Let X be a finite, simple graph without loop, on n vertices, with eigenvalues $\mu_0 \geq \mu_1 \geq \cdots \geq \mu_{n-1}$. Show that $\sum_{i=0}^{n-1} \mu_i = 0$, that $\sum_{i=0}^{n-1} \mu_i^2$ is twice the number of edges in X, and that $\sum_{i=0}^{n-1} \mu_i^3$ is six times the number of triangles in X.

7. Let $X = (V, E)$ be a graph, not necessarily finite. We say that X has bounded degree if there exists $N \in \mathbb{N}$, such that, for every $x \in V$, one has $\sum_{y \in V} A_{xy} \leq N$. Show that in this case, for any $f \in \ell^2(V)$, one has

$$\|Af\|_2 = \left(\sum_{x \in V} |(Af)(x)|^2\right)^{1/2} \leq N \cdot \|f\|_2 = N \cdot \left(\sum_{x \in V} |f(x)|^2\right)^{1/2};$$

that is, A is a bounded linear operator on the Hilbert space $\ell^2(V)$ (Hint: use the Cauchy–Schwarz inequality.)

1.2. Inequalities on the Spectral Gap

Let $X = (V, E)$ be a graph. For $F \subseteq V$, we define the *boundary* ∂F of F to be the set of edges with one extremity in F and the other in $V - F$. In other words, ∂F is the set of edges connecting F to $V - F$. Note that $\partial F = \partial(V - F)$.

1.2.1. Definition. The *isoperimetric constant*, or *expanding constant* of the graph X, is

$$h(X) = \inf\left\{\frac{|\partial F|}{\min\{|F|, |V - F|\}} : F \subseteq V, 0 < |F| < +\infty\right\}.$$

Note that, if X is finite on n vertices, this can be rephrased as $h(X) = \min\left\{\frac{|\partial F|}{|F|} : F \subseteq V, 0 < |F| \leq \frac{n}{2}\right\}$.

1.2.2. Definition. Let $(X_m)_{m \geq 1}$ be a family of finite, connected, k-regular graphs with $|V_m| \to +\infty$ as $m \to +\infty$. We say that $(X_m)_{m \geq 1}$ is a *family of expanders* if there exists $\varepsilon > 0$, such that $h(X_m) \geq \varepsilon$ for every $m \geq 1$.

1.2.3. Theorem. Let $X = (V, E)$ be a finite, connected, k-regular graph without loops. Let μ_1 be the first nontrivial eigenvalue of X (as in section 1.1). Then

$$\frac{k - \mu_1}{2} \leq h(X) \leq \sqrt{2k \, (k - \mu_1)}.$$

Proof. (a) We begin with the first inequality. We endow the set E of edges with an arbitrarily chosen orientation, allowing one to associate, to any edge $e \in E$, its origin e^- and its extremity e^+. This allows us to define the *simplicial coboundary operator* $d : \ell^2(V) \to \ell^2(E)$, where, for $f \in \ell^2(V)$ and $e \in E$,

$$df(e) = f(e^+) - f(e^-).$$

Endow $\ell^2(V)$ with the hermitian scalar product

$$\langle f \mid g \rangle = \sum_{x \in V} \overline{f(x)} \, g(x)$$

and $\ell^2(E)$ with the analogous one. So we may define the adjoint (or conjugate-transpose) operator $d^* : \ell^2(E) \to \ell^2(V)$, characterized by $\langle df \mid g \rangle = \langle f \mid d^*g \rangle$ for every $f \in \ell^2(V)$, $g \in \ell^2(E)$. Define a function $\delta : V \times E \to \{-1, 0, 1\}$ by

$$\delta(x, e) = \begin{cases} 1 & \text{if } x = e^+ \\ -1 & \text{if } x = e^- \\ 0 & \text{otherwise.} \end{cases}$$

Then one checks easily that, for $e \in E$ and $f \in \ell^2(V)$,

$$df(e) = \sum_{x \in V} \delta(x, e) \, f(x) \,;$$

while, for $v \in V$ and $g \in \ell^2(E)$,

$$d^*g(x) = \sum_{e \in E} \delta(x, e) \, g(e).$$

We then define the *combinatorial Laplace operator* $\Delta = d^*d : \ell^2(V) \to \ell^2(V)$. It is easy to check that

$$\Delta = k \cdot \mathrm{Id} - A \,;$$

in particular, Δ does not depend on the choice of the orientation. For an orthonormal basis of eigenfunctions of A, the operator Δ takes the form

$$\Delta = \begin{pmatrix} 0 & & & \\ & k - \mu_1 & & \bigcirc \\ & & \ddots & \\ \bigcirc & & & k - \mu_{n-1} \end{pmatrix},$$

the eigenvalue 0 corresponding to the constant functions on V. Therefore, if f is a function on V with $\sum_{x \in V} f(x) = 0$ (i.e., f is orthogonal to the constant functions in $\ell^2(V)$), we have

$$\|df\|_2^2 = \langle df \mid df \rangle = \langle \Delta f \mid f \rangle \geq (k - \mu_1) \|f\|_2^2.$$

We apply this to a carefully chosen function f. Fix a subset F of V and set

$$f(x) = \begin{cases} |V - F| & \text{if } x \in F \\ -|F| & \text{if } x \in V - F. \end{cases}$$

Then $\sum_{x \in V} f(x) = 0$ and $\|f\|_2^2 = |F||V - F|^2 + |V - F||F|^2 = |F| \, |V - F||V|$. Moreover,

$$df(e) = \begin{cases} 0 \text{ if } e \text{ connects two vertices either in } F \text{ or in } V - F; \\ \pm |V| \text{ if } e \text{ connects a vertex in } F \text{ with a vertex in } V - F. \end{cases}$$

Hence, $\|df\|_2^2 = |V|^2 \, |\partial F|$. So the previous inequality gives

$$|V|^2 |\partial F| \geq (k - \mu_1) |F||V - F||V|.$$

Hence,

$$\frac{|\partial F|}{|F|} \geq (k - \mu_1) \frac{|V - F|}{|V|}.$$

If we assume $|F| \leq \frac{|V|}{2}$, we get $\frac{|\partial F|}{|F|} \geq \frac{k - \mu_1}{2}$; hence, by definition, $h(X) \geq \frac{k - \mu_1}{2}$.

(b) We now turn to the second inequality, which is more involved. Fix a nonnegative function f on V, and set

$$B_f = \sum_{e \in E} |f(e^+)^2 - f(e^-)^2|.$$

Denote by $\beta_r > \beta_{r-1} > \cdots > \beta_1 > \beta_0$ the values of f, and set

$$L_i = \{x \in V : f(x) \geq \beta_i\} \qquad (i = 0, 1, \ldots, r).$$

Note that $L_0 = V$. (Hence, $\partial L_0 = \emptyset$.) To have a better intuition of what is happening, consider the following example on C_8, the cycle graph with eight vertices.

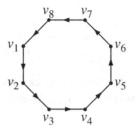

with $f(v_1) = f(v_5) = 4$, $f(v_2) = f(v_6) = f(v_7) = 1$, $f(v_3) = 2$, $f(v_4) = f(v_8) = 3$, so that $\beta_3 = 4 > \beta_2 = 3 > \beta_1 = 2 > \beta_0 = 1$. Then

$$
\begin{aligned}
L_0 &= \{v_1, v_2, v_3, v_4, v_5, v_6, v_7, v_8\}; \\
L_1 &= \{v_1, v_3, v_4, v_5, v_8\}; \\
L_2 &= \{v_1, v_4, v_5, v_8\}; \\
L_3 &= \{v_1, v_5\}; \\
\partial L_0 &= \emptyset; \\
\partial L_1 &= \{\{v_1, v_2\}, \{v_2, v_3\}, \{v_5, v_6\}, \{v_7, v_8\}\}; \ |\partial L_1| = 4; \\
\partial L_2 &= \{\{v_1, v_2\}, \{v_3, v_4\}, \{v_5, v_6\}, \{v_7, v_8\}\}; \ |\partial L_2| = 4; \\
\partial L_3 &= \{\{v_1, v_2\}, \{v_4, v_5\}, \{v_5, v_6\}, \{v_8, v_1\}\}; \ |\partial L_3| = 4.
\end{aligned}
$$

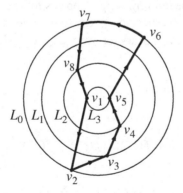

Geometrically, one can envision the graph broken into level curves as follows: L_0 consists of all vertices on or inside the outer-level curve corresponding to $\beta_0 = 1$; L_1 consists of all vertices on or inside the level curve corresponding to $\beta_1 = 2$; and so forth. Then any ∂L_i consists of those edges that reach "downward" from inside L_i to a vertex with a lower value. From the diagram we see clearly that, for example, $\partial L_2 = \{\{v_1, v_2\}, \{v_3, v_4\}, \{v_5, v_6\}, \{v_7, v_8\}\}$.

Coming back to the general case, we now prove the following result about the number B_f.

First Step. $B_f = \sum\limits_{i=1}^{r} |\partial L_i| (\beta_i^2 - \beta_{i-1}^2).$

To see this, we denote by E_f the set of edges $e \in E$, such that $f(e^+) \neq f(e^-)$. Clearly $B_f = \sum\limits_{e \in E_f} |f(e^+)^2 - f(e^-)^2|$. Now, an edge $e \in E_f$ connects some vertex x with $f(x) = \beta_{i(e)}$ to some vertex y with $f(y) = \beta_{j(e)}$. We index these two index values so that $i(e) > j(e)$. Therefore,

$$B_f = \sum_{e \in E_f} (\beta_{i(e)}^2 - \beta_{j(e)}^2)$$

$$= \sum_{e \in E_f} (\beta_{i(e)}^2 - \beta_{i(e)-1}^2 + \beta_{i(e)-1}^2 - \cdots - \beta_{j(e)+1}^2 + \beta_{j(e)+1}^2 - \beta_{j(e)}^2)$$

$$= \sum_{e \in E_f} \sum_{\ell=j(e)+1}^{i(e)} (\beta_\ell^2 - \beta_{\ell-1}^2).$$

Referring to the diagram of level curves, we see that as a given edge e connects a vertex x, with $f(x) = \beta_{i(e)}$, to a vertex y with $f(y) = \beta_{j(e)}$, it crosses every level curve β_ℓ between those two. In the expression for B_f, this corresponds to expanding the term $\beta_{i(e)}^2 - \beta_{j(e)}^2$ by inserting the zero difference $-\beta_\ell^2 + \beta_\ell^2$ for each level curve β_ℓ crossed by the edge e. This means that, in the previous summation for B_f, the term $\beta_\ell^2 - \beta_{\ell-1}^2$ appears for every edge e connecting some vertex x with $f(x) = \beta_i$ and $i \geq \ell$ to some vertex y with $f(y) = \beta_j$ and $j < \ell$. In other words, it appears for every edge $e \in \partial L_\ell$, which establishes the first step.

Second Step. $B_f \leq \sqrt{2k}\, \|df\|_2\, \|f\|_2.$

Indeed,

$$B_f = \sum_{e \in E} |f(e^+) + f(e^-)| \cdot |f(e^+) - f(e^-)|$$

$$\leq \left[\sum_{e \in E} (f(e^+) + f(e^-))^2 \right]^{1/2} \left[\sum_{e \in E} (f(e^+) - f(e^-))^2 \right]^{1/2}$$

$$\leq \sqrt{2} \left[\sum_{e \in E} (f(e^+)^2 + f(e^-)^2) \right]^{1/2} \|df\|_2$$

$$= \sqrt{2k} \left[\sum_{x \in V} f(x)^2 \right]^{1/2} \|df\|_2 = \sqrt{2k}\, \|f\|_2\, \|df\|_2$$

by the Cauchy–Schwarz inequality and the elementary fact that $(a + b)^2 \leq 2(a^2 + b^2)$.

Third Step. Recall that the *support* of f is supp $f = \{x \in V : f(x) \neq 0\}$. Assume that $|\text{supp } f| \leq \frac{|V|}{2}$. Then, $B_f \geq h(X) \|f\|_2^2$.

To see this, notice that $\beta_0 = 0$ and that $|L_i| \leq \frac{|V|}{2}$ for $i = 1, \ldots, r$, so that $|\partial L_i| \geq h(X) |L_i|$ by definition of $h(X)$. So it follows from the first step that

$$B_f \geq h(X) \sum_{i=1}^{r} |L_i| (\beta_i^2 - \beta_{i-1}^2)$$

$$= h(X) \left[|L_r| \beta_r^2 + (|L_{r-1}| - |L_r|) \beta_{r-1}^2 + \cdots + (|L_1| - |L_2|) \beta_1^2 \right]$$

$$= h(X) \left[|L_r| \beta_r^2 + \sum_{i=1}^{r-1} |L_i - L_{i+1}| \beta_i^2 \right];$$

however, since $L_i - L_{i+1}$ is exactly the level set where f takes the value β_i, the term in brackets is exactly $\|f\|_2^2$.

Coda. We now apply this to a carefully chosen function f. Let g be a real-valued eigenfunction for Δ, associated with the eigenvalue $k - \mu_1$. Set $V^+ = \{x \in V : g(x) > 0\}$ and $f = \max \{g, 0\}$. By replacing g by $-g$ if necessary, we may assume $|V^+| \leq \frac{|V|}{2}$. (Note that $V^+ \neq \emptyset$ because $\sum_{x \in V} g(x) = 0$ and $g \neq 0$.) For $x \in V^+$, we have (since $g \leq 0$ on $V - V^+$)

$$(\Delta f)(x) = kf(x) - \sum_{y \in V} A_{xy} f(y) = kg(x) - \sum_{y \in V^+} A_{xy} g(y)$$

$$\leq kg(x) - \sum_{y \in V} A_{xy} g(y) = (\Delta g)(x) = (k - \mu_1) g(x).$$

Using this pointwise estimate, we get

$$\|df\|_2^2 = \langle \Delta f \mid f \rangle = \sum_{x \in V^+} (\Delta f)(x) g(x) \leq (k - \mu_1) \sum_{x \in V^+} g(x)^2$$

$$\leq (k - \mu_1) \|f\|_2^2.$$

Combining the second and third steps, we get

$$h(X) \|f\|_2^2 \leq B_f \leq \sqrt{2k} \|df\|_2 \|f\|_2 \leq \sqrt{2k (k - \mu_1)} \|f\|_2^2,$$

and the result follows by cancelling out. $\square$

From Definition 1.2.2 and Theorem 1.2.3, we immediately deduce the following:

1.2.4. Corollary. Let $(X_m)_{m \geq 1}$ be a family of finite, connected, k-regular graphs without loops, such that $|V_m| \to +\infty$ as $m \to +\infty$. The family $(X_m)_{m \geq 1}$ is a family of expanders if and only if there exists $\varepsilon > 0$, such that $k - \mu_1(X_m) \geq \varepsilon$ for every $m \geq 1$.

This is the spectral characterization of families of expanders: a family of k-regular graphs is a family of expanders if and only if the spectral gap is bounded away from zero. Moreover, it follows from Theorem 1.2.3 that, the bigger the spectral gap, the better "the quality" of the expander.

Exercises on Section 1.2

1. How was the assumption "X has no loop" used in the proof of Theorem 1.2.3?

2. Let X be a finite graph without loop. Choose an orientation on the edges; let d, d^* and $\Delta = d^*d$ be the operators defined in this section. Check that, for $f \in \ell^2(V)$, $x \in V$,

$$\Delta f(x) = \deg(x) \, f(x) - (Af)(x),$$

where $\deg(x)$ is the *degree* of x, i.e., the number of neighboring vertices of x.

3. Using the example given for a function f on the cycle graph C_8, verify that B_f satisfies the first two steps in the proof of the second inequality of Theorem 1.2.3.

4. Show that the multiplicity of the eigenvalue $\mu_0 = K$ is the number of connected components of X.

1.3. Asymptotic Behavior of Eigenvalues in Families of Expanders

We have seen in Corollary 1.2.4 that the quality of a family of expanders can be measured by a lower bound on the spectral gap. However, it turns out that, asymptotically, the spectral gap cannot be too large. All the graphs in this section are supposed to be without loops.

1.3.1. Theorem. Let $(X_m)_{m \geq 1}$ be a family of connected, k-regular, finite graphs, with $|V_m| \to +\infty$ as $m \to +\infty$. Then,

$$\liminf_{m \to +\infty} \mu_1(X_m) \geq 2\sqrt{k-1}.$$

A stronger result will actually be proved in section 1.4. There is an asymptotic threshold, analogous to Theorem 1.3.1, concerning the bottom of the spectrum. Before stating it, we need an important definition.

1.3.2. Definition. The *girth* of a connected graph X, denoted by $g(X)$, is the length of the shortest circuit in X. We will say that $g(X) = +\infty$ if X has no circuit, that is, if X is a tree.

For a finite, connected, k-regular graph, let $\mu(X)$ be the smallest nontrivial eigenvalue of X.

1.3.3. Theorem. Let $(X_m)_{m \geq 1}$ be a family of connected, k-regular, finite graphs, with $g(X_m) \to +\infty$ as $m \to +\infty$. Then

$$\limsup_{m \to +\infty} \mu(X_m) \leq -2\sqrt{k-1}.$$

Theorems 1.3.1 and 1.3.3 single out an extremal condition on finite k-regular graphs, leading to the main definition.

1.3.4. Definition. A finite, connected, k-regular graph X is a *Ramanujan graph* if, for every nontrivial eigenvalue μ of X, one has $|\mu| \leq 2\sqrt{k-1}$.

Assume that $(X_m)_{m \geq 1}$ is a family of k-regular Ramanujan graphs without loop, such that $|V_m| \to +\infty$ as $m \to +\infty$. Then the X_m's achieve the biggest possible spectral gap, providing a family of expanders which is optimal from the spectral point of view.

All known constructions of infinite families of Ramanujan graphs involve deep results from number theory and/or algebraic geometry. As explained in the Overview, our purpose in this book is to give, for every odd prime p, a construction of a family of $(p + 1)$-regular Ramanujan graphs. The original proof that these graphs satisfy the relevant spectral estimates, due to Lubotzky-Phillips, and Sarnak [42], appealed to Ramanujan's conjecture on coefficients of modular forms with weight 2: this explains the chosen terminology. Note that Ramanujan's conjecture was established by Eichler [23].

Exercises on Section 1.3

1. A tree is a connected graph without loops. Show that a k-regular tree T_k must be infinite and that it exists and is unique up to graph isomorphism.

2. Let X be a finite k-regular graph. Fix a vertex x_0 and, for $r < \frac{g(X)}{2}$, consider the ball centered at x_0 and of radius r in X. Show that it is isometric to any ball with the same radius in the k-regular tree T_k. Compute the cardinality of such a ball.

3. Deduce that, if $(X_m)_{m \geq 1}$ is a family of connected k-regular graphs, such that $|V_m| \to +\infty$ as $m \to +\infty$, then

    $$g(X_m) \leq (2 + o(1)) \log_{k-1} |V_m|,$$

    where $o(1)$ is a quantity tending to 0 as $m \to +\infty$.

4. Show that, if $k \geq 5$, one has actually, in exercise 3,

    $$g(X_m) \leq 2 + 2 \log_{k-1} |V_m|.$$

1.4. Proof of the Asymptotic Behavior

In this section we prove a stronger result than that stated in Theorem 1.3.1.

The source of the inequality in Theorem 1.3.1 is the fact that the number of paths of length m from a vertex v to v, in a k-regular graph, is at least the number of such paths from v to v in a k-regular tree. To refine this observation, we count paths without backtracking, and to do this we introduce certain polynomials in the adjacency operator.

Let $X = (V, E)$ be a k-regular, simple graph, with $|V|$ possibly infinite. Recall that we defined a path in X in the Overview. We refine that definition now. A path of length r *without backtracking* in X is a sequence

$$\underline{e} = (x_0, x_1, \ldots, x_r)$$

of vertices in V such that x_i is adjacent to x_{i+1} $(i = 0, \ldots, r - 1)$ and $x_{i+1} \neq x_{i-1}$ $(i = 1, \ldots, r - 1)$. The origin of $\underline{e}$ is x_0, the extremity of $\underline{e}$ is x_r. We define, for $r \in \mathbb{N}$, matrices A_r indexed by $V \times V$, which generalize the adjacency matrix and which are polynomials in A:

$$(A_r)_{xy} = \text{number of paths of length } r, \text{ without backtracking,}$$
$$\text{with origin } x \text{ and extremity } y.$$

Note that $A_0 = \text{Id}$ and that $A_1 = A$, the adjacency matrix. The relationship between A_r and A is the following:

1.4.1. Lemma.

(a) $A_1^2 = A_2 + k \cdot \text{Id}$.

(b) For $r \geq 2$, $A_1 A_r = A_r A_1 = A_{r+1} + (k - 1) A_{r-1}$.

Proof.

(a) For $x, y \in V$, the entry $(A_1^2)_{xy}$ is the number of all paths of length 2 between x and y. If $x \neq y$, such paths cannot have backtracking; hence, $(A_1^2)_{xy} = (A_2)_{xy}$. If $x = y$, we count the number of paths of length 2 from x to x, and, since X is simple, $(A_1^2)_{xx} = k$.

(b) Let us prove that $A_r A_1 = A_{r+1} + (k-1) A_{r-1}$ for $r \geq 2$. For $x, y \in V$, the entry $(A_r A_1)_{xy}$ is the number of paths $(x_0 = x, x_1, \ldots, x_r,$ $x_{r+1} = y)$ of length $r+1$ between x and y, without backtracking except possibly on the last step (i.e., $(x_0, x_1, \ldots, x_r)$ has no backtracking). We partition the set of such paths into two classes according to the value of x_{r-1}:
   - if $x_{r-1} \neq y$, then the path $(x_0, \ldots, x_{r+1})$ has no backtracking, and there are $(A_{r+1})_{xy}$ such paths;
   - if $x_{r-1} = y$, then there is backtracking at the last step, and there are $(k-1)(A_{r-1})_{xy}$ such paths.

We leave the proof of $A_1 A_r = A_{r+1} + (k-1) A_{r-1}$ as an exercise. $\square$

From Lemma 1.4.1, we can compute the *generating function* of the A_r's, that is, the formal power series with coefficients A_r. It turns out to have a particularly nice expression; namely, we have the following:

1.4.2. Lemma.

$$\sum_{r=0}^{\infty} A_r t^r = \frac{1 - t^2}{1 - At + (k-1)t^2}.$$

(This must be understood as follows: in the ring $\mathrm{End}\, \ell^2(V)[[t]]$ of formal power series over $\mathrm{End}\, \ell^2(V)$, we have

$$\left(\sum_{r=0}^{\infty} A_r t^r \right) (\mathrm{Id} - At + (k-1)t^2 \,\mathrm{Id}) = (1 - t^2)\,\mathrm{Id}.)$$

Proof. This is an easy check using Lemma 1.4.1. $\square$

In order to eliminate the numerator $1 - t^2$ in the right-hand side of 1.4.2, we introduce polynomials T_m in A given by

$$T_m = \sum_{0 \leq r \leq \frac{m}{2}} A_{m-2r} \qquad (m \in \mathbb{N}).$$

The generating function of the T_m's is readily computed.

1.4.3. Lemma.

$$\sum_{m=0}^{\infty} T_m t^m = \frac{1}{1 - At + (k-1)t^2}.$$

Proof.

$$\sum_{m=0}^{\infty} T_m t^m = \sum_{m=0}^{\infty} \sum_{0 \le r \le \frac{m}{2}} A_{m-2r} t^m = \sum_{r=0}^{\infty} \sum_{m \ge 2r} A_{m-2r} t^m$$

$$= \sum_{r=0}^{\infty} t^{2r} \sum_{m \ge 2r} A_{m-2r} t^{m-2r} = \left(\sum_{r=0}^{\infty} t^{2r} \right) \left(\sum_{\ell=0}^{\infty} A_\ell t^\ell \right)$$

$$= \frac{1}{1 - t^2} \cdot \frac{1 - t^2}{1 - At + (k-1)t^2} = \frac{1}{1 - At + (k-1)t^2}$$

by Lemma 1.4.2.   □

1.4.4. Definition.

The *Chebyshev polynomials of the second kind* are defined by expressing $\frac{\sin(m+1)\theta}{\sin\theta}$ as a polynomial of degree m in $\cos\theta$:

$$U_m(\cos\theta) = \frac{\sin(m+1)\theta}{\sin\theta} \qquad (m \in \mathbb{N}).$$

For example, $U_0(x) = 1$, $U_1(x) = 2x$, $U_2(x) = 4x^2 - 1, \ldots$. Using trigonometric identities, we see that these polynomials satisfy the following recurrence relation:

$$U_{m+1}(x) = 2x\, U_m(x) - U_{m-1}(x).$$

As in Lemma 1.4.2, from this recurrence relation, we compute the generating function of the U_m's; namely,

$$\sum_{m=0}^{\infty} U_m(x) t^m = \frac{1}{1 - 2xt + t^2}.$$

Performing a simple change of variables, we then compute the generating function of the related family of polynomials $(k-1)^{\frac{m}{2}} U_m\left(\frac{x}{2\sqrt{k-1}} \right)$:

$$\sum_{m=0}^{\infty} (k-1)^{\frac{m}{2}} U_m\left(\frac{x}{2\sqrt{k-1}} \right) t^m = \frac{1}{1 - xt + (k-1)t^2}.$$

In comparison to Lemma 1.4.3, we immediately get the following expression for the operators T_m as polynomials of degree m in the adjacency matrix.

1.4.5. Proposition. For $m \in \mathbb{N}$: $T_m = (k-1)^{\frac{m}{2}} U_m \left(\frac{A}{2\sqrt{k-1}} \right)$. $\square$

Assume that $X = (V, E)$ is a finite, k-regular graph on n vertices, with spectrum

$$\mu_0 = k \geq \mu_1 \geq \cdots \geq \mu_{n-1}.$$

In Proposition 1.4.5, we are going to estimate the trace of T_m in two different ways. This will lead to the trace formula for X.

First, working from a basis of eigenfunctions of A, we have, from Proposition 1.4.5,

$$\mathrm{Tr}\, T_m = (k-1)^{\frac{m}{2}} \sum_{j=0}^{n-1} U_m \left(\frac{\mu_j}{2\sqrt{k-1}} \right).$$

On the other hand, by definition of T_m,

$$\mathrm{Tr}\, T_m = \sum_{0 \leq r \leq \frac{m}{2}} \mathrm{Tr}\, A_{m-2r} = \sum_{x \in V} \sum_{0 \leq r \leq \frac{m}{2}} (A_{m-2r})_{xx}.$$

For $x \in V$, denote by $f_{\ell,x}$ the number of paths of length ℓ in X, without backtracking, with origin and extremity x; in other words, $f_{\ell,x} = (A_\ell)_{xx}$. Then we get the trace formula:

1.4.6. Theorem.

$$\sum_{x \in V} \sum_{0 \leq r \leq \frac{m}{2}} f_{m-2r,x} = (k-1)^{\frac{m}{2}} \sum_{j=0}^{n-1} U_m \left(\frac{\mu_j}{2\sqrt{k-1}} \right),$$

for every $m \in \mathbb{N}$.

We say that X is *vertex-transitive* if the group Aut X of automorphisms of X acts transitively on the vertex-set V. Specifically, this means that for every pair of vertices x and y, there exists $\alpha \in$ Aut X, such that $\alpha(x) = y$. Under this assumption, the number $f_{\ell,x}$ does not depend on the vertex x, and we denote it simply by f_ℓ.

1.4.7. Corollary. Let X be a vertex-transitive, finite, k-regular graph on n vertices. Then, for every $m \in \mathbb{N}$,

$$n \cdot \sum_{0 \leq r \leq \frac{m}{2}} f_{m-2r} = (k-1)^{\frac{m}{2}} \sum_{j=0}^{n-1} U_m \left(\frac{\mu_j}{2\sqrt{k-1}} \right). \quad \square$$

The value of the trace formula 1.4.6 is the following: only looking at the right-hand side (called the spectral side) $(k-1)^{\frac{m}{2}} \sum_{j=0}^{n-1} U_m \left(\frac{\mu_j}{2\sqrt{k-1}} \right)$, it is not obvious that it defines a nonnegative integer. As we shall now explain, the mere positivity of the spectral side has nontrivial consequences. We first need a somewhat technical result about the Chebyshev polynomials.

1.4.8. Proposition. Let $L \geq 2$ and $\varepsilon > 0$ be real numbers. There exists a constant $C = C(\varepsilon, L) > 0$ with the following property: for any probability measure ν on $[-L, L]$, such that $\int_{-L}^{L} U_m \left(\frac{x}{2} \right) d\nu(x) \geq 0$ for every $m \in \mathbb{N}$, we have

$$\nu \left[2 - \varepsilon, L \right] \geq C.$$

(Thus, ν gives a measure at least C to the interval $[2 - \varepsilon, L]$.)

Proof. It is convenient to introduce the polynomials $X_m(x) = U_m \left(\frac{x}{2} \right)$; they satisfy $X_m(2 \cos \theta) = \frac{\sin(m+1)\theta}{\sin \theta}$ and the recursion formula $X_{m+1}(x) = x X_m(x) - X_{m-1}(x)$. It is clear from the first relation that the roots of X_m are $2 \cos \frac{\ell \pi}{m+1}$ $(\ell = 1, \ldots, m)$. In particular the largest root of X_m is $\alpha_m = 2 \cos \frac{\pi}{m+1}$. The proof is then in several steps.

First Step. For $k \leq \ell$: $X_k X_\ell = \sum_{m=0}^{k} X_{k+\ell-2m}$.

We prove this by induction over k. Since $X_0(x) = 1$ and $X_1(x) = x$, the formula is obvious for $k = 0, 1$. (For $k = 1$, this is nothing but the recursion formula.) Then, for $k \geq 2$, we have, by induction hypothesis,

$$\begin{aligned}
X_k X_\ell &= (x X_{k-1} - X_{k-2}) X_\ell \\
&= x \left(X_{k+\ell-1} + X_{k+\ell-3} + \cdots + X_{\ell-k+3} + X_{\ell-k+1} \right) \\
&\quad - \left(X_{k+\ell-2} + X_{k+\ell-4} + \cdots + X_{\ell-k+4} + X_{\ell-k+2} \right) \\
&= \left(X_{k+\ell} + X_{k+\ell-2} \right) + \left(X_{k+\ell-2} + X_{k+\ell-4} \right) \\
&\quad + \cdots + \left(X_{\ell-k+4} + X_{\ell-k+2} \right) + \left(X_{\ell-k+2} + X_{\ell-k} \right) \\
&\quad - \left(X_{k+\ell-2} + X_{k+\ell-4} + \cdots + X_{\ell-k+4} + X_{\ell-k+2} \right) \\
&= X_{k+\ell} + X_{k+\ell-2} + \cdots + X_{\ell-k+2} + X_{\ell-k}.
\end{aligned}$$

Second Step.

$$\frac{X_m(x)}{x - \alpha_m} = \sum_{i=0}^{m-1} X_{m-1-i}(\alpha_m) \cdot X_i(x).$$

Indeed,

$$(x - \alpha_m) \left(\sum_{i=0}^{m-1} X_{m-1-i}(\alpha_m) X_i(x) \right)$$

$$= X_{m-1}(\alpha_m) X_1(x) + \sum_{i=1}^{m-1} X_{m-1-i}(\alpha_m)(X_{i+1}(x) + X_{i-1}(x))$$

$$- \sum_{i=0}^{m-1} X_{m-1-i}(\alpha_m) \alpha_m X_i(x)$$

$$= (X_{m-2}(\alpha_m) - X_{m-1}(\alpha_m) \alpha_m) X_0(x)$$

$$+ \sum_{i=1}^{m-2} (X_{m-i}(\alpha_m) + X_{m-i-2}(\alpha_m) - \alpha_m X_{m-1-i}(\alpha_m)) X_i(x)$$

$$+ (X_1(\alpha_m) - \alpha_m X_0(\alpha_m)) X_{m-1}(x) + X_0(\alpha_m) X_m(x).$$

Now $X_0(\alpha_m) = 1$ and $X_1(\alpha_m) - \alpha_m X_0(\alpha_m) = 0$; in the summation $\sum_{i=1}^{m-2}$ all the coefficients are 0, by the recursion formula. Finally, $X_{m-2}(\alpha_m) - X_{m-1}(\alpha_m) \alpha_m = -X_m(\alpha_m) = 0$, by definition of α_m.

Third Step. Set $Y_m(x) = \frac{X_m(x)^2}{x - \alpha_m}$; then $Y_m = \sum_{i=0}^{2m-1} y_i X_i$, with $y_i \geq 0$.

Indeed, by the second step we have $Y_m = \sum_{i=0}^{m-1} X_{m-1-i}(\alpha_m) X_i X_m$. Now observe that the sequence $\alpha_m = 2 \cos \frac{\pi}{m+1}$ increases to 2. So for $j < m$: $X_j(\alpha_m) > 0$ (since $\alpha_m > \alpha_j$ and α_j is the largest root of X_j). This means that all coefficients are positive in the previous formula for Y_m. By the first step, each $X_i X_m$ is a linear combination, with nonnegative coefficients, of $X_0, X_1, \ldots, X_{2m-1}$, so the result follows.

Fourth Step. Fix $\varepsilon > 0$, $L \geq 2$. For every probability measure ν on $[-L, L]$ such that $\int_{-L}^{L} X_m(x) \, d\nu(x) \geq 0$ for every $m \in \mathbb{N}$, we have $\nu [2 - \varepsilon, L] > 0$.

Indeed, assume by contradiction that $\nu [2 - \varepsilon, L] = 0$; i.e. the support of ν is contained in $[-L, 2 - \varepsilon]$. Take m large enough to have $\alpha_m > 2 - \varepsilon$. Since $Y_m(x) \leq 0$ for $x \leq \alpha_m$, we then have $\int_{-L}^{L} Y_m(x) \, d\nu(x) \leq 0$. On the other hand, by the third step and the assumption on v, we clearly have $\int_{-L}^{L} Y_m(x)\,dv(x) \geq 0$. So $\int_{-L}^{L} Y_m(x)\,dv(x) = 0$, which implies that v is supported in the finite set F_m of zeroes of Y_m; as before we have $F_m = \{2\cos\frac{\ell\pi}{m+1} : 1 \leq \ell \leq m\}$. But this holds for every m large enough. And clearly, since $m+1$ and $m+2$ are relatively prime, we have $F_m \cap F_{m+1} = \emptyset$, so that supp v is empty. But this is absurd.

Coda. Fix $\varepsilon > 0$, $L \geq 2$. Let f be the continuous function on $[-L, L]$ defined by

$$f(x) = \begin{cases} 0 & \text{if } x \leq 2 - \varepsilon \\ 1 & \text{if } x \geq 2 - \frac{\varepsilon}{2} \\ \frac{2}{\varepsilon}(x - 2 + \varepsilon) & \text{if } 2 - \varepsilon \leq x \leq 2 - \frac{\varepsilon}{2}. \end{cases}$$

On $\left[2 - \varepsilon, 2 - \frac{\varepsilon}{2}\right]$, the function f linearly interpolates between 0 and 1. For every probability measure v on $[-L, L]$, we then have

$$v\,[2 - \varepsilon, L] \geq \int_{-L}^{L} f(x)\,dv(x) \geq v\left[2 - \frac{\varepsilon}{2}, L\right].$$

Let $\wp$ be the set of probability measures v on $[-L, L]$, such that $\int_{-L}^{L} X_m(x)\,dv(x) \geq 0$ for every $m \geq 1$. For $v \in \wp$, we have by the fourth step $\int_{-L}^{L} f(x)\,dv(x) > 0$. But $\wp$ is compact in the weak topology and, since f is continuous, the map

$$\wp \to \mathbb{R}^+ : v \mapsto \int_{-L}^{L} f(x)\,dv(x)$$

is weakly continuous. By compactness there exists $C(\varepsilon, L) > 0$, such that $\int_{-L}^{L} f(x)\,dv(x) \geq C(\varepsilon, L)$ for every $v \in \wp$. A fortiori $v\,[2 - \varepsilon, L] \geq C(\varepsilon, L)$, and the proof is complete. (Note that, in the final step, the need for introducing the function f comes from the fact that the map $\wp \to \mathbb{R}^+ : v \mapsto v\,[2 - \varepsilon, L]$ is, *a priori*, not weakly continuous; however, it is bounded below by a continuous function, to which the compactness argument applies.) $\square$

Coming back to the spectra of finite connected, k-regular graphs, we now reach the promised improvement of Theorem 1.3.1: it shows not only that the first nontrivial eigenvalue becomes asymptotically larger than $2\sqrt{k-1}$, but also that a positive proportion of eigenvalues lies in any interval $\left[(2 - \varepsilon)\sqrt{k-1}, k\right]$.

1.4.9. Theorem. For every $\varepsilon > 0$, there exists a constant $C = C(\varepsilon, k) > 0$, such that, for every connected, finite, k-regular graph X on n vertices, the number of eigenvalues of X in the interval $\left[(2 - \varepsilon)\sqrt{k-1}, k\right]$ is at least $C \cdot n$.

Proof. Take $L = \frac{k}{\sqrt{k-1}} \geq 2$ and $\nu = \frac{1}{n} \sum_{j=0}^{n-1} \delta_{\frac{\mu_j}{\sqrt{k-1}}}$ (where δ_a is the Dirac measure at $a \in [-L, L]$, that is, the probability measure on $[-L, L]$ such that $\int_{-L}^{L} f(x) \, d\,\delta_a(x) = f(a)$, for every continuous function f on $[-L, L]$). Then ν is a probability measure on $[-L, L]$, and $\int_{-L}^{L} U_m \left(\frac{x}{2} \right) d\nu(x) = \frac{1}{n} \sum_{j=0}^{n-1} U_m \left(\frac{\mu_j}{2\sqrt{k-1}} \right)$ is nonnegative, by the trace formula 1.4.6. So the assumptions of Proposition 1.4.8 are satisfied, and therefore there exists $C = C(\varepsilon, k) > 0$ such that $\nu \, [2 - \varepsilon, L] \geq C$. But

$$\nu \, [2 - \varepsilon, L] = \frac{1}{n} \times (\text{number of } j\text{'s with } 2 - \varepsilon \leq \frac{\mu_j}{\sqrt{k-1}} \leq L)$$

$$= \frac{1}{n} \times (\text{number of eigenvalues of } X \text{ in } [(2 - \varepsilon)\sqrt{k-1}, \, k]).$$

$\square$

Continuing this analysis we prove the following:

1.4.10. Theorem. Let $(X_m)_{m \geq 1}$ be a sequence of connected, k-regular, finite graphs for which $g(X_m) \to \infty$ as $m \to \infty$. If $\nu_m = \nu(X_m)$ is the measure on $\left[-\frac{k}{\sqrt{k-1}}, \frac{k}{\sqrt{k-1}} \right]$ defined by

$$\nu_m = \frac{1}{|X_m|} \sum_{j=0}^{|X_m|-1} \frac{\delta_{\mu_j}(X_m)}{\sqrt{k-1}},$$

then, for every continuous function f on $\left[-\frac{k}{\sqrt{k-1}}, \frac{k}{\sqrt{k-1}} \right]$,

$$\lim_{m \to \infty} \int_{\frac{-k}{\sqrt{k-1}}}^{\frac{k}{\sqrt{k-1}}} f(x) \, d\nu_m(x) = \int_{-2}^{2} f(x) \sqrt{4 - x^2} \, \frac{dx}{2\pi}.$$

In other words, the sequence of measures $(\nu_m)_{m \geq 1}$ on $\left[-\frac{k}{\sqrt{k-1}}, \frac{k}{\sqrt{k-1}} \right]$ weakly converges to the measure ν supported on $[-2, 2]$, given by $d\nu(x) = \frac{\sqrt{4-x^2}}{2\pi} \, dx$.

Proof. Set $L = \frac{k}{\sqrt{k-1}}$. Recall that $f_{\ell,x}$ denotes the number of paths of length ℓ, without backtracking, from x to x in X_m. We have that for $n \geq 1$, fixed and m large enough (precisely $g(X_m) > n$):

$$f_{n-2r,x} = 0$$

for any $x \in X_m$ and $0 \leq r \leq \frac{n}{2}$. Hence, for m large enough the left-hand side of the equation in Theorem 1.4.6 is zero. Thus, so is the right-hand side, and therefore

$$\int_{-L}^{L} U_n \left(\frac{x}{2}\right) dv_m(x) = 0.$$

We also have that

$$\int_{-L}^{L} U_0 \left(\frac{x}{2}\right) dv_m(x) = 1.$$

For $n \geq 0$, let us compute $\int_{-L}^{L} U_n \left(\frac{x}{2}\right) dv(x)$, using the change of variables $x = 2\cos\theta$:

$$\int_{-L}^{L} U_n \left(\frac{x}{2}\right) dv(x) = \int_{0}^{\pi} U_n(\cos\theta) \, 2\sin^2\theta \, \frac{d\theta}{\pi}$$

$$= \frac{1}{\pi} \int_{0}^{\pi} 2\sin((n+1)\theta) \sin\theta \, d\theta$$

$$= \delta_{n,0}.$$

Hence, for any $n \geq 0$,

$$\lim_{m \to \infty} \int_{-L}^{L} U_n \left(\frac{x}{2}\right) dv_m(x) = \int_{-L}^{L} U_n \left(\frac{x}{2}\right) dv(x).$$

From the recursion relation following Definition 1.4.4, it is clear that the linear span of $U_0 \left(\frac{x}{2}\right), U_1 \left(\frac{x}{2}\right), \ldots, U_n \left(\frac{x}{2}\right)$ is equal to the space of polynomials of degree at most n. Hence we have that

$$\lim_{m \to \infty} \int_{-L}^{L} p(x) \, dv_m(x) = \int_{-L}^{L} p(x) \, dv(x)$$

for any polynomial $p(x)$. The rest of the argument is a standard $\frac{\varepsilon}{3}$ reasoning: fix a continuous function f on $[-L, L]$, and a positive number $\varepsilon > 0$. By the Weierstrass approximation theorem, find a polynomial p such that

$$|f(x) - p(x)| \leq \varepsilon$$

for every $x \in [-L, L]$. Then

$$\left| \int_{-L}^{L} f(x) \, dv_m(x) - \int_{-L}^{L} f(x) \, dv(x) \right|$$

$$\leq \left| \int_{-L}^{L} (f(x) - p(x)) \, dv_m(x) \right| + \left| \int_{-L}^{L} p(x) \, dv_m(x) - \int_{-L}^{L} p(x) \, dv(x) \right|$$

$$+ \left| \int_{-L}^{L} (p(x) - f(x)) \, dv(x) \right|.$$

Since ν_m and ν are probability measures, the first and last term are less than $\frac{\varepsilon}{3}$, while the second term is less than $\frac{\varepsilon}{3}$ for m large enough. So

$$\left| \int_{-L}^{L} f(x)\, d\nu_m(x) - \int_{-L}^{L} f(x)\, d\nu(x) \right| \leq \varepsilon,$$

for m large, which concludes the proof. $\square$

We can now prove the following result, analogous to Theorem 1.4.9, which improves on Theorem 1.3.3.

1.4.11. Corollary. Let $(X_m)_{m \geq 1}$ be a family of connected, k-regular, finite graphs, with $g(X_m) \to \infty$ as $m \to \infty$. For every $\varepsilon > 0$, there exists a constant $C => 0$, such that the number of eigenvalues of X_m in the interval $[-k, (-2+\varepsilon)\sqrt{k-1}]$ is at least $C\,|X_m|$.

Proof. The proof is similar to the last step of the proof of Theorem 1.4.8. We use the same notation as that in Theorem 1.4.10. Let f be the function which is 1 on $\left[-\frac{k}{\sqrt{k-1}}, -2\right]$, 0 on $\left[-2+\varepsilon, \frac{k}{\sqrt{k-1}}\right]$, and interpolates linearly between 1 and 0 on $[-2, -2+\varepsilon]$. Then, for every $m \geq 1$,

$$\nu_m\left[-\frac{k}{\sqrt{k-1}}, -2+\varepsilon\right] \geq \int_{-\frac{k}{\sqrt{k-1}}}^{\frac{k}{\sqrt{k-1}}} f(x)\, d\nu_m(x).$$

For $m \to \infty$, using Theorem 1.4.10, this gives

$$\liminf_{m \to \infty} \nu_m\left[-\frac{k}{\sqrt{k-1}}, (-2+\varepsilon)\right] \geq \int_{-2}^{2} f(x)\, d\nu(x).$$

In other words,

$$\liminf_{m \to \infty} \frac{1}{|X_m|} \times \{\text{number of eigenvalues of } X_m \text{ in } [-k, (-2, \varepsilon)\sqrt{k-1}]\}$$

$$\geq \int_{-2}^{-2+\varepsilon} f(x)\, d\nu(x),$$

from which the result follows. $\square$

Exercises on Section 1.4

1. Complete the proof of Lemma 1.4.1 and prove Lemma 1.4.2.

2. Establish the recursion formula for the Chebyshev polynomials of the second kind, and compute their generating functions.

3. Fix real numbers $L \geq 2$ and $\varepsilon > 0$. Let M be the set of probability measures on $[-L, L]$, endowed with the weak topology. Show that the function $M \to \mathbb{R}^+ : \nu \mapsto \nu[2 - \varepsilon, L]$ is not weakly continuous.

4. (Do not try this exercise if you have never heard about representations of $SU(2)$.) Let Π_m be the unique irreducible representation of $SU(2)$ on $\mathbb{C}^{m+1}$. Set $t_\theta = \begin{pmatrix} e^{i\theta} & 0 \\ 0 & e^{-i\theta} \end{pmatrix} \in SU(2)$. Show that $\operatorname{Tr} \Pi_m(t_\theta) = U_m(\cos \theta)$. Use the Clebsch–Gordan formula to give an alternate proof of the first step in the proof of Proposition 1.4.8.

1.5. Independence Number and Chromatic Number

Let $X = (V, E)$ be a finite graph without loop; as usual we denote by A the adjacency matrix of X.

1.5.1. Definition.

(a) The *chromatic number* $\chi(X)$ is the minimal number of classes in a partition $V = V_1 \cup V_2 \cup \cdots \cup V_\chi$, such that, for every $i = 1, \ldots, \chi$ and every $x, y \in V_i$, we have $A_{xy} = 0$. (In other words, this is the minimal number of colors necessary to paint the vertices of X in such a way that two adjacent vertices have different colors.)

(b) The *independence number* $i(X)$ is the maximal cardinality of a subset $F \subseteq V$, such that $A_{xy} = 0$ for every $x, y \in F$.

These two quantities are related by the following inequality:

1.5.2. Lemma. Let X be a finite graph without loop, on n vertices. Then $n \leq i(X)\chi(X)$.

Proof. Let $V = V_1 \cup V_2 \cup \cdots \cup V_{\chi(X)}$ be a coloring of V in $\chi(X)$ colors. Since $|V_i| \leq i(X)$ for $i = 1, \ldots, \chi(X)$, we have $n = \sum_{i=1}^{\chi(X)} |V_i| \leq i(X) \chi(X)$. $\square$

For a finite, connected, k-regular graph with spectrum

$$k = \mu_0 > \mu_1 \geq \cdots \geq \mu_{n-1},$$

we can relate $i(X)$ to the spectrum of X.

1.5.3. Proposition. Let X be a finite, connected, k-regular graph on n vertices. Then $i(X) \leq \frac{n}{k} \max\{|\mu_1|, |\mu_{n-1}|\}$.

Proof. Let $F \subseteq V$ be a subset of V, of cardinality $|F| = i(X)$, such that $A_{xy} = 0$ for $x, y \in F$. As in the first part of the proof of Theorem 1.2.3, we consider the function $f \in \ell^2(V)$, defined by

$$f(x) = \begin{cases} |V - F| & \text{if } x \in F; \\ -|F| & \text{if } x \in V - F. \end{cases}$$

Then $\sum_{x \in V} f(x) = 0$ and $\|f\|_2^2 = |F| \cdot |V - F| \cdot |V| \leq i(X) n^2$. Take $x \in F$; since $A_{xy} = 0$ for $y \in F$, we have

$$(Af)(x) = \sum_{y \notin F} A_{xy} f(y) = -|F| \sum_{y \notin F} A_{xy} = -|F| \sum_{y \in V} A_{xy} = -ki(X),$$

so that $\|Af\|_2^2 \geq \sum_{x \in F} (Af)(x)^2 = k^2 i(X)^3$.

In an orthonormal basis of eigenfunctions, A takes the form

$$A = \begin{pmatrix} k & & & \\ & \mu_1 & & \bigcirc \\ & & \ddots & \\ & \bigcirc & & \mu_{n-1} \end{pmatrix}.$$

Since $\sum_{x \in V} f(x) = 0$, we have $\|Af\|_2 \leq \max\{|\mu_1|, |\mu_{n-1}|\} \cdot \|f\|_2$. Using the lower bound for $\|Af\|_2$ and the upper bound for $\|f\|_2$, we get

$$k i(X)^{3/2} \leq \max\{|\mu_1|, |\mu_{n-1}|\} \cdot n \cdot i(X)^{1/2},$$

cancelling out $i(X)^{1/2}$. The result follows. $\square$

From Lemma 1.5.2, Propositions 1.5.3 and 1.1.4, and Definition 1.3.4, we immediately get:

1.5.4. Corollary. Let X be a finite, connected, k-regular graph on n vertices, without loop. Then

$$\chi(X) \geq \frac{k}{\max\{|\mu_1|, |\mu_{n-1}|\}}.$$

Moreover, if X is a nonbipartite Ramanujan graph, then

$$\chi(X) \geq \frac{k}{2\sqrt{k-1}} \sim \frac{\sqrt{k}}{2}. \qquad \square$$

Exercises on Section 1.5

1. What do the results of this section become for bipartite graphs?

2. For the complete graph K_n and the cycle graph C_n, compute the chromatic and independence numbers and verify Lemma 1.5.2 and Proposition 1.5.3.

1.6. Large Girth and Large Chromatic Number

A combinatorial problem that has attracted much attention is to construct graphs with large chromatic number and large girth. Note that adding edges increases (or at least does not decrease) the chromatic number but that it does decrease the girth. Given this tension, it is by no means obvious that such graphs exist.

A method, known as the probabilistic method, and due to Erdös [24], has proven to be very powerful in demonstrating the existence of such combinatorial objects. One proceeds by examining the graphs of a certain shape which do not satisfy the desired properties and by showing that these are relatively rare. In this way, most objects (i.e., the "random object") have the desired property and, in particular, their existence is assured. Of course such an argument offers no clue as to be able to find, or give, explicit examples. (These will be reached in section 4.4.)

Let k and c be given large numbers. We seek a graph X with $g(X) \geq k$ and $\chi(X) \geq c$. Let n be an integer which will go to infinity in the following discussion. Consider the set of all graphs on n labeled vertices which have m edges. We denote this set by $\mathcal{X}_{n,m}$. Fix ε such that $0 < \varepsilon < \frac{1}{k}$; set $m = [n^{1+\varepsilon}]$, where [ ] denotes the integer part.

First Step. We start by counting the number of elements in $\mathcal{X}_{n,m}$. To construct a graph $X \in \mathcal{X}_{n,m}$, we must select m edges out of the $\binom{n}{2}$ possible edges. So $|\mathcal{X}_{n,m}| = \binom{\binom{n}{2}}{m}$.

Second Step. We are interested in those X's in $\mathcal{X}_{n,m}$ with small independence number (and hence, by Lemma 1.5.2, large chromatic number). Take η with $0 < \eta < \frac{\varepsilon}{2}$, and set $p = \left[n^{1-\eta} \right]$. To formalize smallness of independence number, we will first say that, for every subset with p elements in the vertex set, the graph X meets the complete graph K_p on these p vertices, in a "large" number of edges, say at least n edges. So we count as "bad" X's the ones which meet a given complete graph K_p (on our vertex set) in few edges. The number of such X's which meet a given K_p in exactly $0 \leq \ell \leq n$ edges is clearly

$$\binom{\binom{p}{2}}{\ell} \binom{\binom{n}{2} - \binom{p}{2}}{m - \ell}.$$

Thus, the number $\widetilde{N}(n, m)$ of $X \in \mathcal{X}_{n,m}$ which meet this given K_p in at most n edges is

$$\widetilde{N}(n, m) = \sum_{\ell=0}^{n} \binom{\binom{p}{2}}{\ell} \binom{\binom{n}{2} - \binom{p}{2}}{m - \ell}.$$

For $n \leq \frac{N}{2}$ and $0 \leq \ell \leq n$, we have

$$\binom{N}{\ell} \leq \binom{N}{n}$$

(see exercise 1). So, for n large and $0 \leq \ell \leq n$, we estimate $\binom{\binom{p}{2}}{\ell} \leq \binom{\binom{p}{2}}{n}$ and $\binom{\binom{n}{2} - \binom{p}{2}}{m - \ell} \leq \binom{\binom{n}{2} - \binom{p}{2}}{m}$. Thus,

$$\begin{aligned}
\widetilde{N}(n, m) &\leq (n+1) \binom{\binom{p}{2}}{n} \binom{\binom{n}{2} - \binom{p}{2}}{m} \leq p^{2n} \binom{\binom{n}{2} - \binom{p}{2}}{m} \\
&= \frac{p^{2n}}{m!} \left[\binom{n}{2} - \binom{p}{2}\right]\left[\binom{n}{2} - \binom{p}{2} - 1\right] \ldots \left[\binom{n}{2} - \binom{p}{2} - m + 1\right].
\end{aligned}$$

Now, for $0 \leq \ell \leq m$, we have

$$\binom{n}{2} - \binom{p}{2} - \ell \leq \left(\binom{n}{2} - \ell\right)\left(1 - \frac{\binom{p}{2}}{\binom{n}{2}}\right),$$

so that

$$\begin{aligned}
\widetilde{N}(n, m) &\leq \frac{p^{2n}}{m!} \binom{n}{2}\left[\binom{n}{2} - 1\right] \ldots \left[\binom{n}{2} - m + 1\right]\left(1 - \frac{\binom{p}{2}}{\binom{n}{2}}\right)^m \\
&= p^{2n} \binom{\binom{n}{2}}{m}\left(1 - \frac{\binom{p}{2}}{\binom{n}{2}}\right)^m \\
&\leq p^{2n} \binom{\binom{n}{2}}{m}\left(1 - \left(\frac{p-1}{n-1}\right)^2\right)^m.
\end{aligned}$$

Now for $0 < x < 1$, we have $(1 - x)^m < e^{-mx}$ hence, by the first step,

$$\widetilde{N}(n, m) \leq p^{2n} e^{-m\left(\frac{p-1}{n-1}\right)^2} |\mathcal{X}_{n,m}|.$$

Third Step. Let $N(n, m)$ be the number of $X \in \mathcal{X}_{n,m}$ which meet *some* K_p in at most n edges. Since the number of possible K_p's is $\binom{n}{p}$, we have

$$N(n, m) \le \binom{n}{p} \tilde{N}(n, m).$$

Fourth Step. Since $\binom{n}{p} \le n^p \le p^n$ (because $p = [n^{1-\eta}]$), we have, by the second and third steps,

$$N(n, m) \le p^{3n} e^{-m\left(\frac{p-1}{n-1}\right)^2} |\mathcal{X}_{n,m}|.$$

Fifth Step. Recall that $0 < \eta < \frac{\varepsilon}{2}$, that $m = [n^{1+\varepsilon}]$, and that $p = [n^{1-\eta}]$. As $n \to \infty$, we have

$$N(n, m) = o\,(|\mathcal{X}_{n,m}|),$$

where the notation $A(n) = o\,(B(n))$ as $n \to \infty$, means $\frac{A(n)}{B(n)} \to 0$ as $n \to \infty$.

Put in another way, this step ensures that the proportion of X's in $\mathcal{X}_{n,m}$ which meet *every* K_p in at least n edges tends to 1 as $n \to \infty$. This will be used to ensure that the independence number is small.

Sixth Step. Next we address the girth. There is no reason that our good X's cited previously have large girth. We will arrange this by removing from X small circuits. Define the integer-valued function F on $\mathcal{X}_{n,m}$ by setting $F(X)$ to be the number of circuits in X of length $\ell \le k$, where k is the large number fixed at the very beginning. Denote by $A(n, k)$ the average value of F:

$$A(n, k) = \frac{1}{|\mathcal{X}_{n,m}|} \sum_{X \in \mathcal{X}_{n,m}} F(X).$$

Seventh Step. We can calculate A another way, that is, by calculating the contribution to the sum of each fixed circuit of length ℓ, say $x_1 \to x_2 \to \ldots \to x_\ell \to x_1$, with $3 \le \ell \le k$. Indeed each such circuit contributes 1 to the sum, for each of $\binom{\binom{n}{2} - \ell}{m - \ell}$ graphs X's. Now there are $n(n-1)\ldots(n-\ell+1)$

such circuits of length ℓ. Hence, we have

$$A(n,k) = \frac{1}{|\mathcal{X}_{n,m}|} \sum_{\ell=3}^{k} n(n-1)\ldots(n-\ell+1) \left(\begin{array}{c} \binom{n}{2} - \ell \\ m - \ell \end{array} \right)$$

$$\leq \sum_{\ell=3}^{k} n^{\ell} \frac{\left(\begin{array}{c} \binom{n}{2} - \ell \\ m - \ell \end{array} \right)}{\left(\begin{array}{c} \binom{n}{2} \\ m \end{array} \right)} \qquad \text{(by the first step)}$$

$$= \sum_{\ell=3}^{k} n^{\ell} \frac{m(m-1)\ldots(m-\ell+1)}{\binom{n}{2}(\binom{n}{2}-1)\ldots(\binom{n}{2}-\ell+1)} \leq \sum_{\ell=3}^{k} \frac{n^{\ell} m^{\ell}}{\binom{n}{2}(\binom{n}{2}-1)\ldots(\binom{n}{2}-\ell+1)}$$

$$= \sum_{\ell=3}^{k} \frac{n^{\ell} m^{\ell}}{\binom{n}{2}^{\ell}} \left[1 + \left(\frac{\binom{n}{2}^{\ell}}{\binom{n}{2}(\binom{n}{2}-1)\ldots(\binom{n}{2}-\ell+1)} - 1 \right) \right].$$

The term in parentheses is a $o(1)$, as $n \to \infty$. This gives the estimate

$$A(n,k) \leq (1+o(1)) \sum_{\ell=3}^{k} \frac{n^{\ell} m^{\ell}}{\binom{n}{2}^{\ell}} = (1+o(1)) \sum_{\ell=3}^{k} \left(\frac{2m}{n-1} \right)^{\ell}$$

$$\leq (1+o(1)) k \cdot \left(\frac{2m}{n-1} \right)^{k} = o(n),$$

since $m = [n^{1+\varepsilon}]$ and $\varepsilon < \frac{1}{k}$.

Eighth Step. It follows that

$$\frac{1}{|\mathcal{X}_{n,m}|} \sum_{X \in \mathcal{X}_{n,m}: F(X) \geq \frac{n}{k}} \frac{n}{k} \leq \frac{1}{|\mathcal{X}_{n,m}|} \sum_{X \in \mathcal{X}_{n,m}} F(X) = A(n,k) = o(n),$$

as $n \to \infty$. Hence,

$$\frac{|\{X \in \mathcal{X}_{n,m} : F(X) \geq \frac{n}{k}\}|}{|\mathcal{X}_{n,m}|} = o(1),$$

as $n \to \infty$.

Coda. For $X \in \mathcal{X}_{n,m}$, consider the two following properties:

1. X meets every K_p in at least n edges;
2. $F(X) < \frac{n}{k}$.

Combining the fifth and eighth steps, we see that, as $n \to \infty$, the proportion of $X \in \mathcal{X}_{n,m}$, which satisfy (1) and (2), tends to 1. So for n large enough (depending on k, ε, η), we choose such an X satisfying (1) and (2). Delete from X all edges which lie on closed circuits of length at most k, getting a graph X'. Clearly $g(X') > k$. Also, according to (2), we have deleted less than n edges in going from X to X'. From (1) it then follows that X' meets every K_p in at least one edge. That is, $i(X') \leq p$. Thus, according to Lemma 1.5.2, we have $\chi(X') \geq \frac{n}{p}$, which is of order n^η and hence is greater than c for n large enough. Thus, X' (which is the "random" modified element of $\mathcal{X}_{n,m}$) fulfills our requirements.

Exercises on Section 1.6

1. Check that, for $0 \leq \ell \leq n \leq \frac{N}{2}$, one has $\binom{N}{\ell} \leq \binom{N}{n}$.

2. According to the construction in section 1.6, how large does n need to be taken in order to have $g(X) \geq 10$ and $\chi(X) \geq 10$?

1.7. Notes on Chapter 1

1.1. The results in section 1.1 can also be derived from the classical Perron–Frobenius theory. For a detailed treatment of the relation between the combinatorics of a graph and the spectrum of its adjacency matrix, see, e.g., the books by Biggs [5] and Chung [15].

1.2. For treatments of families of expanders, see the books by Lubotzky [41] and Sarnak [57]. As indicated in the Overview, the construction of families of expanders is an important problem in network theory. The first constructions go back to the years 1972–73: using counting arguments, Pinsker [52] gave a nonexplicit construction, while Margulis [46] gave an explicit one by appealing to Kazhdan's property (T) in the representation theory of locally compact groups (see [17], Chapter 7, and [41], Chapter 3). A drawback of this second method is that it gives *a priori* no estimate on the size of the ε in Definition 1.2.2 and, therefore, no measure of the quality of the expanders. This problem was first overcome by Gabber and Galil [28]. (See also [2], [16], [25]; as well as recent works by Wigderson & Zuckerman [70].)

The inequalities in Theorem 1.2.3 are often called the Cheeger–Buser inequalities, by analogy with Riemannian geometry. Indeed, in 1970, Cheeger [13] defined the isoperimetric constant of a compact Riemannian manifold M of dimension n:

$$h(M) = \inf \left\{ \frac{\operatorname{vol}_{n-1}(\partial U)}{\min\{\operatorname{vol}_n(U), \operatorname{vol}_n(M-U)\}} \right\},$$

where U runs among nonempty open subsets with smooth boundary ∂U, and vol_n denotes n-dimensional Riemannian volume. He proved that $h(M) \leq 2\sqrt{\lambda_1(M)}$, where $\lambda_1(M)$ is the first nontrivial eigenvalue of the Laplace operator on M. Then, in 1982, Buser [12] proved that $h(M)$ is also bounded by a function of $\lambda_1(M)$:

$$\lambda_1(M) \leq 2a(n-1)h(M) + 10h(M)^2,$$

where the constant $a \geq 0$ is related to the Ricci curvature of M by the inequality $\operatorname{Ricci}(M) \geq -(n-1)a^2$.

The first inequality in Theorem 1.2.3 is due to N. Alon and Milman [3]; the second is due to Dodziuk [22]. The first step in the proof of the second inequality is often called the *co-area formula*, again by analogy with Riemannian geometry: to compute the integral of a function, integrate the volume of the level sets over the range of the function.

1.3. and 1.4. The asymptotic behavior in Theorem 1.3.1 is due to Alon and Boppana (see [42] and [51]); it had several improvements, due to Burger [10], Serre [62], and Grigorchuk and Zuk [31]. We have chosen Serre's approach (Proposition 1.4.8 and Theorem 1.4.9); our proof of Proposition 1.4.8 is a slight improvement of the one in [62]; for explicit estimates of the constant C appearing there, see pp. 213–213 in [39].

The asymptotic behavior, in Theorem 1.3.3, for the bottom of the spectrum is due to Li and Solé [40]. Note that the number $2\sqrt{k-1}$ appearing in Theorems 1.3.1 and 1.3.3 can be understood as follows: let T_k be the k-regular tree: this is the common universal cover of all the finite, connected, k-regular graphs; by exercise 7 of section 1.1, the adjacency matrix A of T_k is a bounded operator on the Hilbert space $\ell^2(V)$ (where V is the set of vertices of T_k). The spectrum of A on $\ell^2(V)$ is then the interval $[-2\sqrt{k-1}, 2\sqrt{k-1}]$, and its spectral measure is essentially the measure ν in Theorem 1.4.10: this is a result of Kesten [36], and actually 1.3.1 can be proved (as in [57]) by "comparing" a finite, connected, k-regular graph to its universal cover and then applying Kesten's result.

As mentioned in the Overview, infinite families of k-regular Ramanujan graphs have been constructed for the following values of k:

- $k = p + 1$, p an odd prime (see [42], [46]);
- $k = 3$ [14];
- $k = q + 1$, q a prime power [48].

The other values of k are *open*, the first open value being $k = 7$.

An intriguing observation is: when one estimates the expanding constant by means of the spectral gap, something is lost in the use of the Cheeger–Buser inequality 1.2.3. Recent results by Brooks and Zuk [9] show that the asymptotic behavior of the expanding constant can be essentially different from the asymptotic behavior of the first nontrivial eigenvalue of the adjacency matrix.

1.5. Proposition 1.5.3 is due to Hoffmann [33].

Chapter 2
Number Theory

2.1. Introduction

The constructions in the later chapters depend on some old results in the theory of numbers. In particular, we need the fact, arguably known to Diophantus around 400 AD and proved first by Lagrange in 1770, that every natural number can be written as a sum of four squares. A remarkable theorem of Jacobi gives an exact formula for the number of representations of n in the form $a_0^2 + a_1^2 + a_2^2 + a_3^2$, in terms of the divisors of n. In section 2.4, we will prove this result for odd n and use it repeatedly later.

We will also need analogous statements about sums of two squares. Since for any integer a, we have $a^2 \equiv 0$ or 1 (mod. 4), it follows that any $n \equiv 3$ (mod. 4) is not a sum of two squares. Still there is an exact formula, due to Legendre, for the number of solutions to $a^2 + b^2 = n$. We prove it in section 2.2 and make use of it as well.

Notice that any sum of two squares can be factored as

$$n = a^2 + b^2 = (a + bi)(a - bi) = \alpha \, \overline{\alpha},$$

where α is an element of the ring of Gaussian integers

$$\mathbb{Z}[i] = \{a + bi : a, b \in \mathbb{Z}, \ i^2 = -1\}.$$

The product $\alpha \, \overline{\alpha}$ is called the norm of α, denoted $N(\alpha)$. Thus, n is a sum of two squares if and only if it is the norm of a Gaussian integer. It turns out to be simpler to work in this ring. We study the arithmetic of $\mathbb{Z}[i]$, extending the familiar notions of integer, prime number, and factorization. The theory for this ring is presented in section 2.2; it is very similar to that for $\mathbb{Z}$.

In a similar way, any representation of a natural number as a sum of four squares can be expressed in terms of norms of elements in yet another ring, the integral quaternions. This ring denoted by $\mathbb{H}(\mathbb{Z})$ is defined by

$$\mathbb{H}(\mathbb{Z}) = \{a_0 + a_1 \, i + a_2 \, j + a_3 \, k : a_0, a_1, a_2, a_3 \in \mathbb{Z},$$
$$i^2 = j^2 = k^2 = -1, \ ij = k, \ jk = i, \ ki = j,$$
$$ji = -k, \ kj = -i, \ ik = -j\}.$$

$\mathbb{H}(\mathbb{Z})$ is not commutative. As with $\mathbb{Z}[i]$ we have conjugate pairs of integral quaternions

$$\alpha = a_0 + a_1 i + a_2 j + a_3 k, \ \overline{\alpha} = a_0 - a_1 i - a_2 j - a_3 k,$$

and the norm $N(\alpha) = \alpha \overline{\alpha} = \overline{\alpha} \alpha = a_0^2 + a_1^2 + a_2^2 + a_3^2$ is multiplicative:

$$N(\alpha \beta) = N(\alpha) N(\beta).$$

Thus, the problem of expressing n as a sum of four squares becomes one of factorization theory in $\mathbb{H}(\mathbb{Z})$. In section 2.6, we therefore study the arithmetic of this ring.

The algebraic structure of the graphs constructed in Chapter 4 depends on the equation $x^2 \equiv p$ (mod. q), where p and q are odd prime numbers. There is a beautiful reciprocity due to Gauss relating the solvability of this equation to the one with p and q reversed. In section 2.3, we give one of the many well-known proofs of this famous "quadratic reciprocity theorem."

2.2. Sums of Two Squares

The study of sums of two squares originated with the problem of Pythagorean triples, or triples (a, b, c) of positive integers, such that $a^2 + b^2 = c^2$. Examples of such triples $3^2 + 4^2 = 5^2$ are $5^2 + 12^2 = 13^2$. The description of Pythagorean triples, as well as the fact that infinite many exist, goes back at least to Diophantus. In the 17th century, Fermat described all integers – not just perfect squares – that could be written as sums of two squares; he and his successors, including Euler, went on to study sums of three or more squares.

The aim of this section is twofold: first, we will prove the Fermat – Euler characterization of integers that are sums of two squares; then, we will show Legendre's formula for the number of representations of a given integer as a sum of two squares.

For $k \geq 2$ and $n \in \mathbb{N}$, we denote by $r_k(n)$ the number of representations of n as a sum of k-squares, that is, the number of solutions of the Diophantine equation $x_0^2 + x_1^2 + \cdots + x_{k-1}^2 = n$:

$$r_k(n) = \left| \left\{ (x_0, \ldots, x_{k-1}) \in \mathbb{Z}^k : \sum_{i=0}^{k-1} x_i^2 = n \right\} \right|.$$

We shall need the *ring of Gaussian integers*,

$$\mathbb{Z}[i] = \{a + bi : a, b \in \mathbb{Z}\},$$

which is easily shown to be a subring of $\mathbb{C}$. For $\alpha = a + bi \in \mathbb{Z}[i]$, we define the *norm* $N(\alpha)$ of α as

$$N(\alpha) = \alpha \bar{\alpha} = |\alpha|^2 = a^2 + b^2.$$

As is customary in algebra, we define the norm *without* taking the square root; in this way, $N(\alpha)$ is a rational integer, as we will now call the integers of $\mathbb{Z}$. Thus, a rational integer is a sum of two squares if and only if it is the norm of some Gaussian integer. A crucial property of the norm is the fact that it is multiplicative:

$$N(\alpha\beta) = N(\alpha) N(\beta) \qquad (\alpha, \beta \in \mathbb{Z}[i]).$$

Note that this immediately shows that products of sums of two squares are sums of two squares. We say that $\alpha \in \mathbb{Z}[i] - \{0\}$ is a *unit* if α is invertible in $\mathbb{Z}[i]$; i.e., $\frac{1}{\alpha} \in \mathbb{Z}[i]$. Since in this case $1 = N\left(\alpha \cdot \frac{1}{\alpha}\right) = N(\alpha) N\left(\frac{1}{\alpha}\right)$, we see that α is a unit in $\mathbb{Z}[i]$ if and only if $N(\alpha) = 1$. Furthermore $N(\alpha) = 1$ if and only if $\alpha \in \{1, -1, i, -i\}$.

2.2.1. Definition.

1. Two Gaussian integers α, β are *associate* if there exists a unit $\varepsilon \in \mathbb{Z}[i]$, such that $\alpha = \varepsilon\beta$.
2. A Gaussian integer $\pi \in \mathbb{Z}[i]$ is *prime* if π is not a unit in $\mathbb{Z}[i]$ and, for any factorization $\pi = \alpha\beta$ in $\mathbb{Z}[i]$, either α or β is a unit in $\mathbb{Z}[i]$.

Note that "being associate" is an equivalence relation on $\mathbb{Z}[i]$, preserving such properties as invertibility, primality, and divisibility. In commutative ring theory, elements satisfying the condition in Definition 2.2.1 (2) are usually called *irreducibles*, while primes are defined by requiring that, if π divides a product, then π divides one of the factors. However, whenever Bézout's relation holds, the two definitions are equivalent. We will show that this is precisely the case for $\mathbb{Z}[i]$ in Propositions 2.2.4 and 2.2.5, but first we begin with the Euclidean algorithm for the Gaussian integers.

2.2.2. Proposition. Let $\alpha, \beta \in \mathbb{Z}[i]$, $\beta \neq 0$. There exists $\gamma, \delta \in \mathbb{Z}[i]$, such that $\alpha = \beta\gamma + \delta$ and $N(\delta) < N(\beta)$.

Proof. Since $\beta \neq 0$, we can form the complex number

$$\frac{\alpha}{\beta} = x + iy \qquad (x, y \in \mathbb{R}).$$

Let $m, n \in \mathbb{Z}$ be such that $|x - m| \leq \frac{1}{2}$ and $|y - n| \leq \frac{1}{2}$. Set $\gamma = m + ni \in \mathbb{Z}[i]$ and $\delta = \beta[(x - m) + i(y - n)]$. Clearly $\frac{\alpha}{\beta} = \gamma + \frac{\delta}{\beta}$; i.e., $\delta = \alpha - \beta\gamma$, so that δ is a Gaussian integer. Finally,

$$\left|\frac{\delta}{\beta}\right|^2 = (x - m)^2 + (y - n)^2 \le \frac{1}{2},$$

so that $N(\delta) \le \frac{1}{2} N(\beta) < N(\beta)$. $\square$

2.2.3. Definition. Fix $\alpha, \beta \in \mathbb{Z}[i]$:

(i) α *divides* β if there exists $\gamma \in \mathbb{Z}[i]$ such that $\beta = \gamma\alpha$;
(ii) $\delta \in \mathbb{Z}[i]$ is a *greatest common divisor* of α and β if δ divides α and β, and whenever $\gamma \in \mathbb{Z}[i]$ divides α and β, it also divides δ.

It is clear that a greatest common divisor, if it exists, is unique up to associate. Furthermore, if $(\alpha, \beta) = \pm 1, \pm i$, we say that α and β are *relatively prime*. Clearly, in that case, we can take $(\alpha, \beta) = 1$.

2.2.4. Proposition. For any $\alpha, \beta \in \mathbb{Z}[i] - \{0\}$, there exists a greatest common divisor $(\alpha, \beta) \in \mathbb{Z}[i]$. Moreover, Bézout's relation holds; that is, there exist $\gamma, \delta \in \mathbb{Z}[i]$ such that $(\alpha, \beta) = \alpha\gamma + \beta\delta$.

Proof. Set $I = \{\alpha\gamma + \beta\delta : \gamma, \delta \in \mathbb{Z}[i]\}$: The reader can easily check that I is closed under addition and subtraction and, further, that if $\lambda \in I$ and $\mu \in \mathbb{Z}[i]$, then $\lambda\mu \in I$. Thus, I forms what is called an *ideal* in $\mathbb{Z}[i]$. Let $\lambda_0 = \alpha\gamma_0 + \beta\delta_0$ be a nonzero element of minimal norm in I. We claim that $(\alpha, \beta) = \lambda_0$. Indeed, by Proposition 2.2.2, we can find $\sigma, \tau \in \mathbb{Z}[i]$, such that $\alpha = \sigma\lambda_0 + \tau$ and $N(\tau) < N(\lambda_0)$. Then $\tau = \alpha - \sigma\lambda_0$ belongs to I and, by the minimality of $N(\lambda_0)$, we must have $\tau = 0$. Hence, λ_0 divides α. Similarly, λ_0 divides β. Since $\lambda_0 = \alpha\gamma_0 + \beta\delta_0$, every common divisor of α and β must divide λ_0. Finally, Bézout's relation holds with $\gamma = \gamma_0, \delta = \delta_0$. $\square$

2.2.5. Proposition. $\pi \in \mathbb{Z}[i]$ is prime if and only if, whenever π divides a product $\alpha\beta$ $(\alpha, \beta \in \mathbb{Z}[i])$, it divides either α or β.

Proof.

($\Rightarrow$) If π divides $\alpha\beta$, we have $\alpha\beta = \pi\sigma$ for some $\sigma \in \mathbb{Z}[i]$. We may assume that π does not divide α and must then show that π divides β. Consider (π, α): since it divides π, which is prime, we must have $(\pi, \alpha) = 1$. Then, by our previous result,

$$1 = \pi\gamma + \alpha\delta$$

for some $\gamma, \delta \in \mathbb{Z}[i]$. Then $\beta = \pi\beta\gamma + \alpha\beta\delta = \pi\beta\gamma + \pi\sigma\delta = \pi(\beta\gamma + \sigma\delta)$, showing that π divides β.

($\Leftarrow$) If $\pi = \alpha\beta$, in particular π divides $\alpha\beta$. Say that π divides β; i.e., $\beta = \pi\gamma$ for some $\gamma \in \mathbb{Z}[i]$. Then $\pi = \alpha\beta = \alpha\pi\gamma$; cancelling out, we get $1 = \alpha\gamma$; i.e., α is a unit. $\square$

From this we get unique factorization in $\mathbb{Z}[i]$.

2.2.6. Proposition. Every nonzero element in $\mathbb{Z}[i]$ is, in a unique way, a product of primes in $\mathbb{Z}[i]$. More precisely, if $\alpha \in \mathbb{Z}[i] - \{0\}$, then $\alpha = \pi_1 \dots \pi_k$ for some primes $\pi_1, \dots, \pi_k$ in $\mathbb{Z}[i]$; and if $\alpha = \pi_1 \dots \pi_k = \sigma_1 \dots \sigma_\ell$ are two factorizations of α into primes, then $k = \ell$ and, after permuting the indices, π_i is associate to σ_i, for $1 \le i \le k$.

Proof. Existence is proved by induction over $N(\alpha)$, the case $N(\alpha) = 1$ (i.e., α is a unit) being trivial. So assume $N(\alpha) > 1$; two cases may happen: if α is prime, there is nothing to prove; if α is not prime, we find a factorization $\alpha = \beta\gamma$, where neither β nor γ is invertible. Then $N(\alpha) = N(\beta)N(\gamma)$ with $N(\beta), N(\gamma) < N(\alpha)$. By induction assumption, β and γ are products of primes in $\mathbb{Z}[i]$; hence, so is α.

To prove uniqueness, assume $\alpha = \pi_1 \dots \pi_k = \sigma_1 \dots \sigma_\ell$ as in the statement. We may assume $k \le \ell$. Since π_1 divides $\sigma_1 \dots \sigma_\ell$, and π_1 is prime, by Proposition 2.2.5, we see that π_1 divides at least one of the σ_i's; say π_1 divides σ_1. Write $\sigma_1 = \varepsilon_1 \pi_1$, with $\varepsilon_1 \in \mathbb{Z}[i]$. Since σ_1 is prime, ε_1 must be a unit. Canceling out π_1 in both factorizations, we get $\pi_2 \dots \pi_k = \varepsilon_1 \sigma_2 \dots \sigma_\ell$. Clearly, we may iterate the process, until we get 1 in the left-hand side. Suppose by contradiction $k < \ell$. Then we get $1 = \varepsilon_1 \dots \varepsilon_k \sigma_{k+1} \dots \sigma_\ell$ and, taking norms, we get a contradiction. So $k = \ell$, which concludes the proof. $\square$

We get a first application of the arithmetic of $\mathbb{Z}[i]$ to sums of two squares, a famous result stated by Fermat around 1640 and proved by Euler in 1793. We denote by $\mathbb{F}_q$ the finite field with q elements and by $\mathbb{F}_q^\times$ the multiplicative group of nonzero elements in $\mathbb{F}_q$.

2.2.7. Theorem. Let p be an odd prime in $\mathbb{N}$. The following are equivalent:

(i) $p \equiv 1 \pmod{4}$;
(ii) -1 is a square in $\mathbb{F}_p$; i.e., the congruence $x^2 \equiv -1 \pmod{p}$ has a solution in $\mathbb{Z}$;
(iii) p is a sum of two squares (so $r_2(p) > 0$).

Proof.

(i) ⇔ (ii) For $y \in \mathbb{F}_p^\times$, define the packet of y as

$$P_y = \{y, -y, y^{-1}, -y^{-1}\}.$$

It is easily checked that the packets do partition $\mathbb{F}_p^\times$. There might be some coincidences within a packet P_y. One cannot have $y = -y$ (since y is invertible and p is odd). But one may have $y = y^{-1}$: this happens exactly when $y = \pm 1$, in which case $P_y = \{1, -1\}$. And one may have $y = -y^{-1}$: this happens exactly when -1 is a square modulo p, in which case the corresponding packet P_y has two elements. To summarize, we constructed a partition of $\mathbb{F}_p^\times$ into classes of four elements, with at most two exceptions having two elements each. Note that the exceptional class P_1 is always present. Therefore, if $p \equiv 1$ (mod. 4), there must be two classes with two elements, so that -1 is a square modulo p; and if $p \equiv 3$ (mod. 4), there must be just one exceptional class, namely, P_1, and -1 is not a square modulo p.

(ii) ⇒ (iii) Suppose that -1 is a square modulo p. So we find $x \in \mathbb{Z}$ such that p divides $x^2 + 1$. Write $x^2 + 1 = (x + i)(x - i)$ in $\mathbb{Z}[i]$ and notice that p does not divide either $x + i$ or $x - i$ in $\mathbb{Z}[i]$. By Proposition 2.2.5, this means that p is not a prime in $\mathbb{Z}[i]$. So there exists a factorization $p = \alpha\beta$ in $\mathbb{Z}[i]$, where neither factor is a unit; therefore, $N(\alpha) > 1$ and $N(\beta) > 1$. Taking norms we get $p^2 = N(p) = N(\alpha)N(\beta)$. This implies $N(\alpha) = N(\beta) = p$, so p is a sum of two squares.

(iii) ⇒ (ii) If p is a sum of two squares, say $p = a^2 + b^2$, then a and b are invertible modulo p. So we find $c \in \mathbb{Z}$, such that $bc \equiv 1$ (mod. p). Then $pc^2 = (ac)^2 + (bc)^2$, and reducing modulo p,

$$0 \equiv (ac)^2 + 1 \qquad (\text{mod. } p),$$

so that -1 is a square modulo p. □

Here is now the promised characterization of integers which are sums of two squares: it is a celebrated result of Fermat and Euler.

2.2.8. Corollary. An integer $n \geq 2$ is a sum of two squares (so $r_2(n) > 0$) if and only if every prime number $p \equiv 3$ (mod. 4) appears with even exponent in the factorization of n into primes.

Proof. Let $n = a^2 + b^2$ be a sum of two squares. Let p be an odd prime dividing n. Let p^k be the highest power of p dividing both a and b; set $x = \frac{a}{p^k}$, $y = \frac{b}{p^k}$; then $\frac{n}{p^{2k}} = x^2 + y^2$.

Suppose that p still divides $\frac{n}{p^{2k}}$. Then, as in the proof that (iii) $\Rightarrow$ (ii) in Theorem 2.2.7, one deduces that -1 is a square modulo p. (This follows from the set that p cannot divide both x and y.) So, $p \equiv 1 \pmod{4}$. By contraposition, if $p \equiv 3 \pmod{4}$, then p cannot divide $\frac{n}{p^{2k}}$ further, showing that p appears with an even exponent in the factorization of n.

We leave the proof of the converse as an exercise. $\square$

Coming back to the arithmetic in $\mathbb{Z}[i]$, we notice a useful criterion for a Gaussian integer to be relatively prime to a rational integer.

2.2.9. Lemma. Let $m \in \mathbb{Z}$, $\alpha \in \mathbb{Z}[i]: (m, \alpha) = 1$ if and only if $(m, N(\alpha)) = 1$.

Proof.

($\Rightarrow$) If $(m, \alpha) = 1$, then by Bézout's relation we can find $\gamma, \delta \in \mathbb{Z}[i]$ such that $1 = \gamma m + \delta \alpha$. Then

$$N(\delta) N(\alpha) = N(1 - \gamma m) = (1 - \gamma m)(1 - \overline{\gamma} m)$$
$$= 1 - (\gamma + \overline{\gamma})m + N(\gamma) m^2,$$

or $N(\delta) N(\alpha) + (\gamma + \overline{\gamma})m - N(\gamma)m^2 = 1$. Note that $\gamma + \overline{\gamma}$ and $N(\gamma)$ belong to $\mathbb{Z}$. So, if $\beta \in \mathbb{Z}[i]$ divides both m and $N(\alpha)$, then it divides 1, showing that β is a unit.

($\Leftarrow$) Assume that $(m, N(\alpha)) = 1$. If $\delta \in \mathbb{Z}[i]$ divides both m and α, then δ divides m and $N(\alpha) = \overline{\alpha} \alpha$. Again, δ must divide 1, so δ is a unit. $\square$

We can now characterize the primes in $\mathbb{Z}[i]$.

2.2.10. Proposition. A Gaussian integer $\pi \in \mathbb{Z}[i]$ is prime if and only if one of the following three mutually exclusive cases occur:

(i) $N(\pi) = 2$ (in this case π is an essociate of $1 + i$; that is, $\pi \in \{1 \pm i, -1 \pm i\}$);

(ii) $N(\pi) = p$, where p is a prime in $\mathbb{Z}$ and $p \equiv 1 \pmod{4}$;

(iii) π is associate to q, where q is a prime in $\mathbb{Z}$, and $q \equiv 3 \pmod{4}$.

Proof. ($\Rightarrow$) Let π be a prime in $\mathbb{Z}[i]$, and let p be a prime in $\mathbb{Z}$ dividing $N(\pi)$. Set $\delta = (p, \pi)$. By Lemma 2.1.9, δ is not a unit. Since π is prime, δ is associate to π, so we may assume $\delta = \pi$. Write then $p = \pi \gamma$, for some

$\gamma \in \mathbb{Z}[i]$. Taking norms we have $p^2 = N(\pi) N(\gamma)$, or $p = \frac{N(\pi)}{p} \cdot N(\gamma)$. Two cases then appear:

(a) $\frac{N(\pi)}{p} = 1$, forcing $N(\pi) = p$. Then p is a sum of two squares, and by Theorem 2.2.7, we have either $p = 2$ or $p \equiv 1$ (mod. 4).

(b) $N(\gamma) = 1$, in which case π is associate to p and p is a prime in $\mathbb{Z}[i]$. Then p is not a sum of two squares and therefore $p \equiv 3$ (mod. 4).

($\Leftarrow$) Observe that, if $N(\pi)$ is prime in $\mathbb{Z}$, then π is prime in $\mathbb{Z}[i]$. Indeed, if $\pi = \alpha\beta$, taking norms we get $N(\pi) = N(\alpha) N(\beta)$, which gives immediately that either α or β is invertible. So, if either $N(\pi) = 2$ or $N(\pi) = p$, with $p \equiv 1$ (mod. 4), then π is prime in $\mathbb{Z}[i]$. On the other hand, if q is prime in $\mathbb{Z}$, $q \equiv 3$ (mod. 4), then q remains prime in $\mathbb{Z}[i]$: indeed, if $q = \alpha\beta$ for $\alpha, \beta \in \mathbb{Z}[i]$, then taking norms we get $q^2 = N(\alpha) N(\beta)$; since q is not a sum of two squares by Theorem 2.2.7, we cannot have $N(\alpha) = N(\beta) = q$. Therefore, either α or β is a unit in $\mathbb{Z}[i]$. $\square$

With this in hand, we now reach Legendre's formula for $r_2(n)$, for which we will need some additional notation. For $n \in \mathbb{N}$ we make the following definitions:

- $d_1(n)$ is the number of divisors of $n \in \mathbb{N}$ which are congruent to 1 modulo 4;
- $d_3(n)$ is the number of divisors of $n \in \mathbb{N}$ which are congruent to 3 modulo 4;
- $d(n)$ is the number of divisors of n.

2.2.11. Theorem. For $n \in \mathbb{N}, n > 0 : r_2(n) = 4 (d_1(n) - d_3(n))$.

Proof. Set $\delta(n) = d_1(n) - d_3(n)$. Assume first that $N \in \mathbb{N}$ is odd and write $N = km$, where

$$k = \prod_{h=1}^{a} p_h^{r_h} \quad (p_h \equiv 1 \text{ (mod. 4)}),$$

$$m = \prod_{j=1}^{b} q_j^{s_j} \quad (q_j \equiv 3 \text{ (mod. 4)}).$$

A divisor of N is congruent to 1 modulo 4 if and only if an even number of q_j's, counting multiplicities, appears in its factorization. From this we deduce

$$\delta(N) = d(k) \delta(m).$$

Claim.

$$\delta(m) = \begin{cases} 0 & \text{if at least one } s_j \text{ is odd} \\ 1 & \text{if all } s_j\text{'s are even, that is, if } m \text{ is a square.} \end{cases}$$

To prove the claim, set $m' = \frac{m}{q_1^{s_1}}$. Note that $\delta(1) = 1$. If s_1 is even, then

$$d_1(m) = \left(\tfrac{s_1}{2} + 1\right) d_1(m') + \tfrac{s_1}{2} d_3(m')$$

$$d_3(m) = \tfrac{s_1}{2} d_1(m') + \left(\tfrac{s_1}{2} + 1\right) d_3(m'),$$

so that $\delta(m) = \delta(m')$. If s_1 is odd, then,

$$d_1(m) = \frac{s_1 + 1}{2} d_1(m') + \frac{s_1 + 1}{2} d_3(m') = d_3(m),$$

so that $\delta(m) = 0$ in this case. This proves the claim.

From the claim, we deduce that

$$\delta(N) = \begin{cases} d(k) & \text{if } m \text{ is a square,} \\ 0 & \text{otherwise.} \end{cases}$$

Now, let $n > 0$ be an integer. Write $n = 2^t N$, with N odd and $N = km$ as before. Note that $\delta(n) = \delta(N)$. If m is not a square, then $\delta(n) = 0$ by the previous, and also $r_2(n) = 0$ by Corollary 2.2.8. So the theorem is true in this case. Assume now that m is a square. Then we know, by Corollary 2.2.8 again, that $r_2(n) > 0$. The idea is, on one hand, to write $n = A^2 + B^2$ and, on the other, to factor n into primes in $\mathbb{Z}[i]$ using unique factorization and the description of primes in $\mathbb{Z}[i]$ from Propositions 2.2.6 and 2.2.10. Equating these, we get

$$n = A^2 + B^2 = (A + iB)(A - iB) = (-i)^t (1 + i)^{2t} \prod_{h=1}^{a} \pi_h^{r_h} \bar{\pi}_h^{r_h} \prod_{j=1}^{b} q_j^{s_j},$$

where $\pi_h \in \mathbb{Z}[i]$ is a prime, such that $N(\pi_h) = p_h$. Now $r_2(n)$ is the number of factorizations of n as $(A + iB)(A - iB)$ in $\mathbb{Z}[i]$. By unique factorization, and the fact that $N(A + iB) = N(A - iB)$, we must have

$$A + iB = u(1 + i)^t \prod_{h=1}^{a} \pi_h^{w_h} \bar{\pi}_h^{u_h} \prod_{j=1}^{b} q_j^{\frac{s_j}{2}}$$

$$A - iB = u'(1 + i)^t \prod_{h=1}^{a} \pi_h^{u_h} \bar{\pi}_h^{w_h} \prod_{j=1}^{b} q_j^{\frac{s_j}{2}},$$

with u, u' units, such that $uu' = (-i)^t$ and $u_h + w_h = r_h$ $(1 \le h \le a)$. The freedom lies in the choice of u and of the u_h's. The number of possible choices for $A + iB$ is therefore

$$4 \prod_{h=1}^{a} (r_h + 1) = 4 d(k) = 4 \delta(N) = 4 d(n). \quad \square$$

Fixing $\varepsilon > 0$, we say that a real quantity $f(n)$, depending on $n \in \mathbb{N}$, is a $0_\varepsilon(n^\varepsilon)$ if there exists a constant $C = C(\varepsilon) > 0$, such that

$$|f(n)| \le Cn^\varepsilon$$

for every $n \in \mathbb{N}$. Using Theorem 2.2.11, we may estimate the order of magnitude of $r_2(n)$.

2.2.12. Corollary. For all $\varepsilon > 0 : r_2(n) = 0_\varepsilon(n^\varepsilon)$.

Proof. From Theorem 2.2.11, $r_2(n) \le 4(d_1(n) + d_3(n)) \le 4 d(n)$. We leave it as an exercise to check that $d(n) = 0_\varepsilon(n^\varepsilon)$. $\quad \square$

There is no simple formula as 2.2.11 for $r_3(n)$. (See the notes at the end of Chapter 2.) Nevertheless, we may estimate the order of magnitude of $r_3(n)$.

2.2.13. Corollary. For all $\varepsilon > 0 : r_3(n) = 0_\varepsilon(n^{\frac{1}{2}+\varepsilon})$.

Proof. We have

$$r_3(n) = \sum_{k=0}^{[\sqrt{n}]} r_2(n - k^2)$$

$$\le C(\varepsilon) \sum_{k=0}^{[\sqrt{n}]} (n - k^2)^\varepsilon \qquad \text{by Corollary 2.2.12}$$

$$\le C(\varepsilon) n^{\frac{1}{2}+\varepsilon}. \quad \square$$

Exercises on Section 2.2

1. Describe an infinite, one-parameter family of solutions $x = f(t)$, $y = g(t)$, $z = h(t)$, where $x^2 + y^2 = z^2$ and $(x, y, z) = 1$.

2. Prove, without appealing to Theorem 2.2.11, that $d_1(n) - d_3(n) \ge 0$.

3. Prove that $d(n) = 0_\varepsilon(n^\varepsilon)$ for every $\varepsilon > 0$.

4. Let m, n be rational integers. Prove that m, n are relatively prime in $\mathbb{Z}[i]$, if an only if m, n are relatively prime in $\mathbb{Z}$.

5. Let $n > 0$ be an integer, such that every prime $p \equiv 3$ (mod. 4) appears with an even exponent in n. Prove that n is a sum of two squares. [Hint: use Theorem 2.2.7 and the fact that 2 is a sum of two squares.]

6. Let p be an odd prime. The aim of this exercise is to give a group-theoretical proof of the fact that -1 is a square modulo p if and only if $p \equiv 1$ (mod. 4). First prove that -1 is a square modulo p if and only if the multiplicative group $\mathbb{F}_p^{\times}$ (of order $p - 1$) contains a subgroup of order 4. Conclude by appealing to the fact that $\mathbb{F}_p^{\times}$ is a cyclic group.

2.3. Quadratic Reciprocity

Let p be an odd prime. Theorem 2.2.7 gives a complete answer to the question, "When is -1 a square modulo p?" Quadratic reciprocity, due to Gauss, deals with the more general question, "When is $m \in \mathbb{Z}$ a square modulo p?"

We begin by defining the *Legendre symbol* $\left(\dfrac{m}{p}\right)$ as

$$\left(\frac{m}{p}\right) = \begin{cases} 0 & \text{if } p \text{ divides } m; \\ 1 & \text{if } p \text{ does not divide } m \text{ and } m \text{ is a square modulo } p; \\ -1 & \text{if } p \text{ does not divide } m \text{ and } m \text{ is not a square modulo } p. \end{cases}$$

Using the fact that the group of squares of $\mathbb{F}_p^{\times}$ has index 2 in $\mathbb{F}_p^{\times}$ (see exercise 2 in the following), one deduces the multiplicative relation

$$\left(\frac{mn}{p}\right) = \left(\frac{m}{p}\right)\left(\frac{n}{p}\right) \qquad (m, n \in \mathbb{Z}).$$

The following lemma has its own interest:

2.3.1. Lemma. For $n \in \mathbb{Z} : n^{\frac{p-1}{2}} \equiv \left(\frac{n}{p}\right)$ (mod. p).

Proof. The result is obvious if n is a multiple of p, so we may assume that p does not divide n. Then $n^{p-1} \equiv 1$ (mod. p), by Fermat's theorem, which means that $n^{\frac{p-1}{2}} \equiv \pm 1$ (mod. p). If n is a square modulo p, say $n \equiv m^2$ (mod. p), then $n^{\frac{p-1}{2}} \equiv m^{p-1} \equiv 1$ (mod. p), again by Fermat's theorem. Now, since $\mathbb{F}_p$ is a field, the equation $x^{\frac{p-1}{2}} = 1$ has at most $\frac{p-1}{2}$ solutions in $\mathbb{F}_p$. But we just checked that each square in $\mathbb{F}_p^{\times}$ provides a solution, and there are $\frac{p-1}{2}$ such squares. In other words, the set of solutions of $x^{\frac{p-1}{2}} = 1$ is exactly the set of squares in $\mathbb{F}_p^{\times}$. Therefore, if n is not a square modulo p, we must have $n^{\frac{p-1}{2}} \equiv -1$ (mod. p). $\square$

From the multiplicativity of the Legendre symbol, we see that, to compute $\left(\frac{m}{p}\right)$ for an arbitrary integer m, it is enough to compute the values of the Legendre symbol for primes in $\mathbb{Z}$ and for -1. The answer to this question is provided by Gauss' celebrated *law of quadratic reciprocity*.

2.3.2. Theorem. Let p be an odd prime. Then

(i) $\left(\frac{-1}{p}\right) = (-1)^{\frac{p-1}{2}}$;

(ii) $\left(\frac{2}{p}\right) = (-1)^{\frac{p^2-1}{8}}$;

(iii) if q is an odd prime, distinct from p : $\left(\frac{q}{p}\right) = (-1)^{\frac{(p-1)(q-1)}{4}} \left(\frac{p}{q}\right)$.

Proof.

(i) This is just a rephrasing of the Fermat – Euler Theorem 2.2.7.

(ii) For every integer $k \in \mathbb{Z}$, there is a unique integer $r \in [-p/2, p/2]$, such that $k \equiv r \pmod{p}$; we call r the *minimal residue* of k. Now we compute the minimal residues of the $\frac{p-1}{2}$ numbers $2, 4, 6, \ldots, p-1$. Assume $p \equiv 1 \pmod{4}$; we get $2; 4; \ldots; \frac{p-1}{2}; -\left(\frac{p-3}{2}\right); -\left(\frac{p-7}{2}\right);$ $\ldots; -1$. Note that, in absolute values, we just get a permutation of the $\frac{p-1}{2}$ numbers $1, 2, \ldots, \frac{p-1}{2}$. Also, $\frac{p-1}{4}$ of the minimal residues are negative. Taking the product of them all, we get

$$(-1)^{\frac{p-1}{4}} \prod_{j=1}^{(p-1)/2} j \equiv \prod_{j=1}^{(p-1)/2} (2j) \pmod{p}$$

$$\equiv 2^{\frac{p-1}{2}} \prod_{j=1}^{(p-1)/2} j \pmod{p}.$$

Cancelling out $\prod_{j=1}^{(p-1)/2} j$, we get

$$2^{\frac{p-1}{2}} \equiv (-1)^{\frac{p-1}{4}} \equiv (-1)^{\frac{p^2-1}{8}} \pmod{p}.$$

By Lemma 2.3.1, we also have $\left(\frac{2}{p}\right) \equiv (-1)^{\frac{p^2-1}{8}}$.

For $p \equiv 3 \pmod{4}$, we proceed in the same way, except that there are now $\frac{p+1}{4}$ negative minimal residues, to the effect that $\left(\frac{2}{p}\right) = (-1)^{\frac{p+1}{4}} = (-1)^{\frac{p^2-1}{8}}$ also in this case.

(iii) We first study some properties of the minimal residues. Let m be an integer, not divisible by p. We consider the minimal residues of the $\frac{p-1}{2}$

numbers $m, 2m, \ldots, \frac{p-1}{2} m$. Since these numbers are pairwise non-congruent modulo p, the minimal residues form a family of $\frac{p-1}{2}$ integers in $[-p/2, p/2]$; among these, we denote by $r_1, \ldots, r_\lambda$ the positive ones; and by $-r'_1, \ldots, -r'_\mu$, the negative ones $\left(\lambda + \mu = \frac{p-1}{2} \right)$.

Claim 1. The family $\{r_1, \ldots, r_\lambda, r'_1, \ldots, r'_\mu\}$ is a rearrangement of $\Big\{ 1, 2, \ldots,$ $\frac{p-1}{2} \Big\}$. Indeed, the r_i's and r'_j's are integers between 1 and $\frac{p-1}{2}$. So it is enough to see that $r_i \neq r'_j$, for every i and j. For exactly one $a \in \left\{ 1, \ldots, \frac{p-1}{2} \right\}$ we have $am \equiv r_i$ (mod. p), and for exactly one $b \in \left\{ 1, \ldots, \frac{p-1}{2} \right\}$ we have $bm \equiv -r'_j$ (mod. p). So if $r_i = r'_j$, we get $(a+b)m \equiv 0$ (mod. p); hence $a + b \equiv 0$ (mod. p), which is impossible since $1 \leq a, b \leq \frac{p-1}{2}$. This proves Claim 1.

We now generalize what we did for $m = 2$ in part (ii) of the proof.

Claim 2. With m, μ as before: $\left(\frac{m}{p} \right) = (-1)^\mu$. Indeed, consider

$$\prod_{j=1}^{(p-1)/2} (mj) = m^{\frac{p-1}{2}} \prod_{j=1}^{(p-1)/2} j \, ;$$

by Claim 1, the numbers $r_1, \ldots, r_\lambda, r'_1, \ldots, r'_\mu$ form a rearragement of 1, 2, $\ldots$, $\frac{p-1}{2}$, so we get

$$m^{\frac{p-1}{2}} \prod_{j=1}^{(p-1)/2} j \equiv (-1)^\mu \prod_{j=1}^{(p-1)/2} j \qquad \text{(mod. } p\text{)}.$$

Cancelling out $\prod_{j=1}^{(p-1)/2} j$, we get $m^{\frac{p-1}{2}} \equiv (-1)^\mu$ (mod. p). By Lemma 2.3.1, we have $m^{\frac{p-1}{2}} \equiv \left(\frac{m}{p} \right)$ (mod. p), so that, finally, $\left(\frac{m}{p} \right) = (-1)^\mu$. This proves Claim 2.

We now define $S(p, q) = \sum_{k=0}^{(p-1)/2} \left[\frac{kq}{p} \right]$.

Claim 3. Let μ be the number of negative minimal residues of the sequence $q, 2q, \ldots, \frac{p-1}{2} \cdot q$. Then $S(p, q)$ has the same parity as μ.

To see this, for $k = 1, \ldots, \frac{p-1}{2}$, write $kq = p\left[\frac{kq}{p}\right] + u_k$, with $u_k \in \{1, \ldots, p-1\}$; note that u_k is nothing but the remainder in the Euclidean division of kq by p. If $u_k < \frac{p}{2}$, then u_k is the minimal residue of qk, so that $u_k = r_i$ for exactly one i; if $u_k > \frac{p}{2}$, then $u_k - p$ is the minimal residue of kq, so that $u_k - p = -r'_j$ for a unique j. Set $R = \sum_{i=1}^{\lambda} r_i$ and $R' = \sum_{j=1}^{\mu} r'_j$, so that $R = \sum_{k:u_k<\frac{p}{2}} u_k$ and $R' = \mu p - \sum_{k:u_k>\frac{p}{2}} u_k$.

Since $r_1, \ldots, r_\lambda, r'_1, \ldots, r'_\mu$ is a rearrangement of $1, \ldots, \frac{p-1}{2}$ (by Claim 1), we have

$$\frac{p^2-1}{8} = \sum_{k=1}^{(p-1)/2} k = R + R' = \mu p + \sum_{k:u_k<\frac{p}{2}} u_k - \sum_{k:u_k>\frac{p}{2}} u_k;$$

that is, $\mu + \sum_{k=1}^{(p-1)/2} u_k \equiv \frac{p^2-1}{8}$ (mod. 2). Now, summing $kq = p\left[\frac{kq}{p}\right] + u_k$ from $k = 1$ to $k = \frac{p-1}{2}$, we get

$$q\frac{p^2-1}{8} = p \cdot S(p, q) + \sum_{k=1}^{(p-1)/2} u_k;$$

hence, $\frac{p^2-1}{8} \equiv S(p, q) + \sum_{k=1}^{(p-1)/2} u_k$ (mod. 2). This immediately gives $S(p, q) \equiv \mu$ (mod. 2), proving Claim 3.

Claim 4. $S(p, q) + S(q, p) = \frac{(p-1)(q-1)}{4}$.

To see this, we consider the rectangle $\left[1, \frac{p-1}{2}\right] \times \left[1, \frac{q-1}{2}\right]$ in $\mathbb{R}^2$. Clearly it contains $\frac{p-1}{2} \cdot \frac{q-1}{2}$ integer points. We are going to count these points in another way, counting first the ones below the line $y = \frac{qx}{p}$ and then the ones above that line. (Note that no integral point from the rectangle lies on the line.)

Clearly the number of integer points under the line is $\sum_{k=1}^{(p-1)/2} \left[\frac{kq}{p}\right]$, while the

number of integer points above the line is $\sum_{\ell=1}^{(q-1)/2} \left[\frac{\ell p}{q}\right]$. This proves Claim 4.

To conclude the proof, we now observe that, with μ as in Claim 3, we have

$$\left(\frac{q}{p}\right) = (-1)^{\mu} \quad \text{(by Claim 2)}$$

$$= (-1)^{S(p,q)} \quad \text{(by Claim 3).}$$

Similarly, we have $\left(\frac{p}{q}\right) = (-1)^{S(q,p)}$. Multiplying together, we get $\left(\frac{p}{q}\right)\left(\frac{q}{p}\right) = (-1)^{S(p,q)+S(q,p)} = (-1)^{\frac{(p-1)(q-1)}{4}}$ by Claim 4. This concludes the proof of Gauss' law of quadratic reciprocity. $\quad\Box$

Exercises on Section 2.3

Let p be an odd prime.

1. Prove that $p^2 - 1$ is divisible by 8.

2. Complete the proof of Theorem 2.3.2 (ii) for $p \equiv 3 \pmod{4}$.

3. Prove that there are $\frac{p-1}{2}$ squares in $\mathbb{F}_p^{\times}$. [Hint: show that the map $\mathbb{F}_p^{\times} \to \mathbb{F}_p^{\times} : x \mapsto x^2$ is a group homomorphism. What is its kernel?]

4. Show that

$$\left(\frac{-3}{p}\right) = \begin{cases} 1 & \text{if } p \equiv 1 \pmod{6}; \\ -1 & \text{if } p \equiv -1 \pmod{6}; \\ 0 & \text{if } p = 3. \end{cases}$$

5. Show that

$$\left(\frac{5}{p}\right) = \begin{cases} 1 & \text{if } p \equiv \pm 1 \pmod{10}; \\ -1 & \text{if } p \equiv \pm 3 \pmod{10}; \\ 0 & \text{if } p = 5. \end{cases}$$

2.4. Sums of Four Squares

The aim of this section is to prove the following classical result, due to Jacobi.

2.4.1. Theorem. Let n be an odd positive integer. Then $r_4(n) = 8 \sum_{d|n} d$. The proof rests on three lemmas.

2.4.2. Lemma. For every $n \in \mathbb{N} : r_4(2n) = r_4(4n)$.

Proof. If $x_0^2 + x_1^2 + x_2^2 + x_3^2 = 4n$, one sees by reducing mod. 4 that either all x_i's are even or they are all odd. Therefore, the change of variables

$$y_0 = \frac{x_0 - x_1}{2}, \ y_1 = \frac{x_0 + x_1}{2}, \ y_2 = \frac{x_2 - x_3}{2}, \ y_3 = \frac{x_2 + x_3}{2}$$

(with inverse $x_0 = y_0 + y_1, x_1 = y_1 - y_0, x_2 = y_2 + y_3, x_3 = y_3 - y_2$) maps an integral solution of $x_0^2 + x_1^2 + x_2^2 + x_3^2 = 4n$ to an integral solution of $y_0^2 + y_1^2 + y_2^2 + y_3^2 = 2n$ and establishes a bijection between the two sets of solutions. $\square$

2.4.3. Lemma. For odd $n \in \mathbb{N} : r_4(2n) = 3\,r_4(n)$.

Proof. If $x_0^2 + x_1^2 + x_2^2 + x_3^2 = 2n$, by reducing mod. 4 we see that exactly two of the x_i's are even, while the other ones are odd. Hence, the integer solutions of $x_0^2 + x_1^2 + x_2^2 + x_3^2 = 2n$ can be partitioned into three classes:

$$\begin{cases} x_0 \equiv x_1 \ (\text{mod. } 2) \\ x_2 \equiv x_3 \ (\text{mod. } 2) \end{cases} ; \ \begin{cases} x_0 \equiv x_2 \ (\text{mod. } 2) \\ x_1 \equiv x_3 \ (\text{mod. } 2) \end{cases} ; \ \begin{cases} x_0 \equiv x_3 \ (\text{mod. } 2) \\ x_1 \equiv x_2 \ (\text{mod. } 2) \end{cases} .$$

By changes of variables similar to the ones used in the proof of Lemma 2.4.2, each of these classes is in bijection with the set of solutions of $y_0^2 + y_1^2 + y_2^2 + y_3^2 = n$. $\square$

For the previous lemma, we need one more notation: for $k \geq 2$ and $n \in \mathbb{N}$, let $N_k(n)$ be the number of representations of n as a sum of k squares of positive odd integers:

$$N_k(n) = \left| \left\{ (x_0, \ldots, x_{k-1}) \in \mathbb{N}^k : \sum_{i=0}^{k-1} x_i^2 = n, \right. \right.$$

$$\left. \left. x_i \equiv 1 \,(\text{mod. } 2), \ 0 \leq i \leq k-1 \right\} \right|.$$

2.4.4. Lemma. For odd $n \in \mathbb{N}$: $N_4(4n) = \sum_{d \mid n} d$.

Proof. Noticing that a sum of four squares is a sum of two sums of two squares, we get the convolution formula

$$N_4(4n) = \sum_{\substack{(r,s):r+s=4n \\ r,s \geq 0}} N_2(r) N_2(s).$$

Since a sum of two odd squares is congruent to 2 modulo 4, we may rewrite this as

$$N_4(4n) = \sum_{\substack{(r,s):r+s=4n \\ r \equiv s \equiv 2 \,(\mathrm{mod.}\,4)}} N_2(r) N_2(s).$$

By Theorem 2.2.11, we have $r_2(k) = 4\,(d_1(k) - d_3(k))$. But r_2 counts positive and negative solutions, while N_2 counts only positive solutions; so

$$N_2(r) = d_1(r) - d_3(r).$$

Now $r \equiv 2$ (mod. 4), so $\frac{r}{2}$ is odd. Divisors of r which contribute to the latter formula are exactly divisors of $\frac{r}{2}$; that is,

$$N_2(r) = \sum_{(a,b):r=2ab} (-1)^{\frac{a-1}{2}} ;$$

similarly,

$$N_2(s) = \sum_{(c,d):s=2cd} (-1)^{\frac{c-1}{2}} = \sum_{(c,d):s=2cd} (-1)^{\frac{1-c}{2}}.$$

(Here a, b, c, d are positive, odd integers.) Hence,

$$N_4(4n) = \sum_{\substack{(a,b,c,d):4n=2ab+2cd \\ a,b,c,d>0,\,\mathrm{odd}}} (-1)^{\frac{a-c}{2}}.$$

Now we perform the change of variables

$$a = x + y; \ c = x - y; \ b = z - t; \ d = z + t,$$

with inverse

$$x = \frac{a+c}{2} ; \ y = \frac{a-c}{2} ; \ z = \frac{b+d}{2} ; \ t = \frac{d-b}{2}.$$

Then

$$ab + cd = 2\,(xz - yt)$$

and

$$N_4(4n) = \sum_{\substack{(x,y,z,t):n=xz-yt \\ |y|<x, |t|<z, \, x\not\equiv y \,(\mathrm{mod.}\,2), z\not\equiv t \,(\mathrm{mod.}\,2)}} (-1)^y.$$

Note that x and z are both positive. We now split this sum in three parts:

$$N_4(4n) = N_+ + N_0 + N_-,$$

according to $y > 0$, $y = 0$, $y < 0$.

Claim 1. $N_+ = N_-$. Indeed the change of variables

$$(x, y, z, t) \mapsto (x, -y, z, -t)$$

establishes a bijection between the 4-tuples contributing to N_+ and the ones contributing to N_-.

Claim 2. $N_+ = 0$. First, let us consider the set Q of 4-tuples contributing to N_+:

$$Q = \{(x, y, z, t) : n = xz - yt \, ; \, 0 < y < x \, ; \, |t| < z \, ; \, x \not\equiv y \,(\mathrm{mod.}\,2) \, ;$$

$$z \not\equiv t \,(\mathrm{mod.}\,2)\}.$$

Now we make a change of variables α defined by

$$x' = 2v(x, y)z - t \, ; \, y' = z \, ; \, z' = y \, ; \, t' = 2v(x, y)y - x,$$

where $v(x, y)$ is the unique positive integer v, such that

$$2v - 1 < \frac{x}{y} < 2v + 1.$$

(Since $x \not\equiv y \,(\mathrm{mod.}\,2)$, the rational number $\frac{x}{y} > 1$ is not an odd integer.) We now study properties of α.

   (i) $\alpha(Q) \subset Q$. This is a cumbersome calculation, but a straightforward one.

   (ii) $\alpha^2 = \mathrm{Id}$. To see this, notice that $v(x, y) = v(x', y')$: this follows since $\frac{x'}{y'} = \frac{2v(x,y)z-t}{z} = 2v(x, y) - \frac{t}{z}$ and $2v(x, y) - 1 < 2v(x, y) - \frac{t}{z} < 2v(x, y) + 1$, since $|t| < z$.

     It is then easy to check that $\alpha^2 = \mathrm{Id}$.

   (iii) If $(x, y, z, t) \in Q$, then $y \not\equiv y' \,(\mathrm{mod.}\,2)$. In particular, α is a fixed point free involution of Q. Since $n = xz - yt$ is odd, $xz \not\equiv yt \,(\mathrm{mod.}\,2)$. Noting that $t \not\equiv z \,(\mathrm{mod.}\,2)$ and $x \not\equiv y \,(\mathrm{mod.}\,2)$, the result for y and y' is immediate.

From (i) and (ii) we see that

$$N^+ = \sum_{(x,y,z,t)\in Q} (-1)^y = \sum_{(x',y',z',t')\in Q} (-1)^{y'}.$$

By (iii), the term associated with (x, y, z, t) is the negative of the one associated with (x', y', z', t'). Hence, $N_+ = -N_+$, and so $N_+ = 0$, proving Claim 2.

From Claims 1 and 2, it remains to prove that $N_0 = \sum_{d|n} d$. But, by definition

$$N_0 = |\{(x, z, t) : n = xz, \ |t| < z, \ z \not\equiv t \ (\text{mod. } 2)\}|.$$

Note that, in such a triple (x, z, t), the integer z must be odd since n is. Now, for a fixed odd integer z, there are exactly z even integers in the interval $[-z, z]$. Hence, $N_0 = \sum_{z|n} z$, completing the proof of Lemma 2.4.4. $\square$

Proof of Theorem 2.4.1. We begin with a

Claim. For n odd, we have $r_4(4n) = 16 \, N_4(4n) + r_4(n)$.

Indeed, if $x_0^2 + x_1^2 + x_2^2 + x_3^2 = 4n$, as in the proof of Lemma 2.4.2, we observe that either all x_i's are even or they are all odd. In the first case, the change of variables $y_i = \frac{x_i}{2}$ $(i = 0, 1, 2, 3)$ provides a bijection between the set of even solutions of $x_0^2 + x_1^2 + x_2^2 + x_3^2 = 4n$ and the set of solutions of $y_0^2 + y_1^2 + y_2^2 + y_3^2 = n$; so there are $r_4(n)$ such solutions. In the second case, there are sixteen $N_4(4n)$ solutions (the coefficient 16 coming from the 2^4 possible choices for the signs of the x_i's). This proves the claim.

Then,

$$3 \, r_4(n) = r_4(2n) \quad \text{(by Lemma 2.4.3)}$$
$$= r_4(4n) \quad \text{(by Lemma 2.4.2)}$$
$$= 16 \, N_4(4n) + r_4(n) \quad \text{(by the previous Claim)}$$
$$= 16 \left(\sum_{d|n} d \right) + r_4(n) \quad \text{(by Lemma 2.4.4)}.$$

Cancelling out gives the statement of Theorem 2.4.1. $\square$

Exercises on Section 2.4

1. In the proof of Lemma 2.4.3, check that the inverse changes of variables map solutions of $y_0^2 + y_1^2 + y_2^2 + y_3^2 = n$ to solutions of $x_0^2 + x_1^2 + x_2^2 + x_3^2 = 2n$ which satisfy the correct parity conditions.
2. Fill in the details in the proof of claim 2, in Lemma 2.3.4.

2.5. Quaternions

In section 2.2, we have seen that there is an algebraic structure underlying sums of two squares: the ring of Gaussian integers. It turns out that there is a similar structure underlying sums of four squares: the ring of integer quaternions. However, the lack of commutativity makes this structure rather more subtle. We split the exposition in two parts: in this section, we present general definitions and properties of quaternions over R, an arbitrary commutative ring with a multiplicative identity (usually called a *commutative ring with unit*), and we postpone to section 2.6 the discussion of the arithmetic of integer quaternions.

2.5.1. Definition. The *Hamilton quaternion algebra* over R, denoted by $\mathbb{H}(R)$, is the associative unital algebra given by the following presentation:

(i) $\mathbb{H}(R)$ is the free R-module over the symbols $1, i, j, k$; that is, $\mathbb{H}(R) = \{a_0 + a_1 i + a_2 j + a_3 k : a_0, a_1, a_2, a_3 \in R\}$;
(ii) 1 is the multiplicative unit;
(iii) $i^2 = j^2 = k^2 = -1$;
(iv) $ij = -ji = k; jk = -kj = i; ki = -ik = j$.

This definition is natural, in the sense that any unital ring homomorphism $R_1 \to R_2$ extends to a unital ring homomorphism $\mathbb{H}(R_1) \to \mathbb{H}(R_2)$ by mapping 1 to 1, i to i, j to j and k to k.

If $q = a_0 + a_1 i + a_2 j + a_3 k$ is a quaternion, its *conjugate* quaternion is $\overline{q} = a_0 - a_1 i - a_2 j - a_3 k$. The *norm* of q is $N(q) = q\overline{q} = \overline{q}q = a_0^2 + a_1^2 + a_2^2 + a_3^2$. Note that the quaternionic norm, like the Gaussian norm, is multiplicative; that is, given $q_1, q_2 \in \mathbb{H}(R)$,

$$N(q_1 q_2) = N(q_1) N(q_2).$$

(Hence, the product of two sums of four squares is itself a sum of four squares. This fact is crucial since it reduces the problem of representing any natural number as a sum of four squares to one of finding such a representation for primes alone.)

We will need to identify certain quaternion algebras with algebras of 2×2 matrices over a field. Again, though we will be interested specifically in the finite field $\mathbb{F}_q$, this identification can be defined over more general fields.

The *characteristic of a ring* is either zero or the smallest positive integer m, such that

$$0 = m \cdot 1 = 1 + 1 + \ldots + 1 \qquad (m \text{ times}).$$

In an integral domain and, therefore, in any field, the characteristic must be zero or a prime number. The rationals $\mathbb{Q}$, the real numbers $\mathbb{R}$, and the complex numbers $\mathbb{C}$ are all fields with characteristic zero; while for any prime power $q = p^\ell$, the finite field $\mathbb{F}_q$ has characteristic p. With this in hand, we have the following:

2.5.2. Proposition. Let K be a field, not of characteristic 2. Assume that there exists $x, y \in K$, such that $x^2 + y^2 + 1 = 0$. Then $\mathbb{H}(K)$ is isomorphic to the algebra $M_2(K)$ of 2-by-2 matrices over K.

Proof. Let $\psi : \mathbb{H}(K) \to M_2(K)$ be defined by

$$\psi(a_0 + a_1 i + a_2 j + a_3 k) = \begin{pmatrix} a_0 + a_1 x + a_3 y & -a_1 y + a_2 + a_3 x \\ -a_1 y - a_2 + a_3 x & a_0 - a_1 x - a_3 y \end{pmatrix}.$$

One checks that $\psi(q_1 q_2) = \psi(q_1) \psi(q_2)$ for $q_1, q_2 \in \mathbb{H}(K)$. Since ψ is a K-linear map between two K-vector spaces of the same dimension 4, to prove that ψ is an isomorphism it is enough to show that ψ is injective. But $\psi(a_0 + a_1 i + a_2 j + a_3 k) = 0$ leads to a 4-by-4 homogeneous linear system in the variables a_0, a_1, a_2, a_3, with determinant

$$\begin{vmatrix} 1 & x & 0 & y \\ 0 & -y & 1 & x \\ 0 & -y & -1 & x \\ 1 & -x & 0 & -y \end{vmatrix} = -4(x^2 + y^2) = 4 \neq 0$$

(since char $K \neq 2$). $\qquad \square$

Proposition 2.5.2 obviously applies not only to algebraically closed fields, but also to, as we shall see, any finite field $\mathbb{F}_q$, where q is an odd prime power.

2.5.3. Proposition. Let q be an odd prime power. There exists $x, y \in \mathbb{F}_q$, such that $x^2 + y^2 + 1 = 0$.

First Proof (*Nonconstructive*). Counting 0, there are $\frac{q+1}{2}$ squares in $\mathbb{F}_q$. Define then

$$A_+ = \{1 + x^2 : x \in \mathbb{F}_q\} \, ; \; A_- = \{-y^2 : y \in \mathbb{F}_q\}.$$

Since $|A_+| = |A_-| = \frac{q+1}{2}$, we have $A_+ \cap A_- \neq \emptyset$, hence, the result.

Second Proof (*Constructive*). Clearly it is enough to prove the result for the prime field $\mathbb{F}_p$ (p an odd prime). If -1 is a square modulo p, take the smallest x in $\{2, \ldots, p - 2\}$, such that $x^2 + 1 = 0$, and $y = 0$. If -1 is not a square modulo p, let a be the largest quadratic residue in $\{1, \ldots, p - 2\}$; then $a + 1$ is not a square modulo p, and therefore $-a - 1$ is a square modulo p. Let x (resp. y) be the smallest element in $\{1, \ldots, p - 2\}$, such that $x^2 \equiv a$ (mod. p) (resp. $y^2 \equiv -a - 1$ (mod. p)). Then $x^2 + y^2 + 1 \equiv 0$ (mod. p). Note that none of these proofs make use of the quadratic reciprocity 2.3.2. $\quad\square$

Exercises on Section 2.5

1. For R a unital commutative ring:
   (a) Check that $\mathbb{H}(R)$ is associative;
   (b) Show that the map $\mathbb{H}(R) \to \mathbb{H}(R) : q \mapsto \overline{q}$ is an anti-automorphism (i.e., $\overline{q_1 q_2} = \overline{q_2}\, \overline{q_1}$ for $q_1, q_2 \in \mathbb{H}(R)$).

2. Let $\psi : \mathbb{H}(K) \to M_2(K)$ be the map defined in the proof of Proposition 2.5.2.
   (a) Check that $\psi(q_1 q_2) = \psi(q_1)\, \psi(q_2)$, for $q_1, q_2 \in \mathbb{H}(K)$;
   (b) for $q \in \mathbb{H}(K)$, show that $\det \psi(q) = N(q)$ and that $\operatorname{Tr} \psi(q) = q + \overline{q}$;
   (c) check that ψ maps "real" quaternions (those with $q = \overline{q}$) to scalar matrices.

3. For $q \in \mathbb{H}(\mathbb{Z})$, prove that the following properties are equivalent
   (a) q is invertible in $\mathbb{H}(\mathbb{Z})$;
   (b) $N(q) = 1$;
   (c) $q \in \{\pm 1, \pm i, \pm j, \pm k\}$.

2.6. The Arithmetic of Integer Quaternions

We now restrict ourselves to $\mathbb{H}(\mathbb{Z})$ and explore some arithmetic properties of this particular ring. Its interest comes from the fact that a rational integer is a sum of four squares if and only if it is the norm of some quaternion in $\mathbb{H}(\mathbb{Z})$. From exercise 3 of the previous section, we have seen that the invertible elements, or units, are $\pm 1, \pm i, \pm j, \pm k$. As in $\mathbb{Z}$ and $\mathbb{Z}[i]$, there is a factorization into primes for any integer quaternion, though in $\mathbb{H}(\mathbb{Z})$ this factorization is no longer unique. We will show that as a noncommutative ring, $\mathbb{H}(\mathbb{Z})$ has a modified Euclidean algorithm and corresponding greatest common right-hand and left-hand divisors that are unique up to associates. We will then see that no rational prime remains prime in $\mathbb{H}(\mathbb{Z})$ but, instead, can be factored into a product of two conjugate prime quaternions. In fact, determining whether a quaternion integer is prime is extremely simple: $\alpha \in \mathbb{H}(\mathbb{Z})$ is prime if and only if $N(\alpha)$ is prime in $\mathbb{Z}$. This presents a contrast to the situation in $\mathbb{Z}[i]$, where any rational prime $q \equiv 3$ (mod. 4) remains prime but has Gaussian norm $N(q) = q^2$.

Let us begin with some definitions:

2.6.1. Definition.

    (a) A quaternion $\alpha \in \mathbb{H}(\mathbb{Z})$ is *odd* (respectively, *even*) if $N(\alpha)$ is an odd (respectively, even) rational integer.

    (b) A quaternion $\alpha \in \mathbb{H}(\mathbb{Z})$ is *prime* if α is not a unit in $\mathbb{H}(\mathbb{Z})$, and if, whenever $\alpha = \beta\gamma$ in $\mathbb{H}(\mathbb{Z})$, then either β or γ is a unit.

    (c) Two quaternions $\alpha, \alpha' \in \mathbb{H}(\mathbb{Z})$ are *associate* if there exist unit quaternions $\varepsilon, \varepsilon' \in \mathbb{H}(\mathbb{Z})$, such that $\alpha' = \varepsilon\alpha\varepsilon'$.

    (d) $\delta \in \mathbb{H}(\mathbb{Z})$ is a *right-hand divisor* of $\alpha \in \mathbb{H}(\mathbb{Z})$ if there is $\gamma \in \mathbb{H}(\mathbb{Z})$, such that $\alpha = \gamma\,\delta$.

Because $N(\varepsilon) = 1$ for any unit ε, "being associate" is an equivalence relation on the elements of $\mathbb{H}(\mathbb{Z})$ that preserve arithmetic properties such as being odd or even, being prime, or being a unit.

Recall that, for $\mathbb{Z}$ and $\mathbb{Z}[i]$, we were able to use Bézout's relation to move from the definition of a prime as an irreducible to the following (Proposition 2.2.5). π is a prime if and only if whenever π divides a product xy; then π divides x or π divides y. However, we clearly cannot adapt this statement to the noncommutative ring $\mathbb{H}(\mathbb{Z})$, since a right divisor of xy cannot generally be a possible right divisor of x. Hence, we will need to proceed without that property for primes in $\mathbb{H}(\mathbb{Z})$. Nevertheless, the definition of primes in 2.6.1(b) immediately gives existence of factorization into prime quaternions.

2.6.2. Proposition. Every quaternion $\alpha \in \mathbb{H}(\mathbb{Z})$ is a product of prime quaternions.

Proof. We proceed by induction over $N(\alpha)$, the case $N(\alpha) = 1$ (i.e., α invertible) being trivial. So assume $N(\alpha) > 1$. If α is prime, there is nothing to prove. Otherwise, we find a factorization $\alpha = \beta\gamma$, where neither β nor γ is invertible in $\mathbb{H}(\mathbb{Z})$. So β, γ satisfy $N(\beta) < N(\alpha)$, $N(\gamma) < N(\alpha)$. By induction hypothesis, β and γ are products of primes, and so is α. $\square$

Note that the factorization in Proposition 2.6.2 is not necessarily unique (not even up to associates); e.g.,

$$13 = (1 + 2i + 2j + 2k)(1 - 2i - 2j - 2k) = (3 + 2i)(3 - 2i)$$

are two genuinely different factorizations of 13 into prime quaternions.

We proceed with the partial Euclidean algorithm, that is, one confined to odd quaternions and multiplication on the right. An analogous result holds for multiplication on the left, but the associated γ_1 and δ_1 are not necessarily the same. We will use this right-hand Euclidean algorithm to construct the greatest common right-hand divisor, but obvious modifications in the proofs lead to equivalent left-hand results.

2.6.3. Lemma. Let α and $\beta \in \mathbb{H}(\mathbb{Z})$, with β odd. There exists γ, $\delta \in \mathbb{H}(\mathbb{Z})$, such that

$$\alpha = \gamma\beta + \delta \qquad \text{and} \qquad N(\delta) < N(\beta).$$

Proof. We begin with a

Claim. Given $\sigma = s_0 + s_1 i + s_2 j + s_3 k \in \mathbb{H}(\mathbb{Z})$, and m an odd positive integer, there exists $\gamma \in \mathbb{H}(\mathbb{Z})$, such that $N(\sigma - \gamma m) < m^2$. Indeed, for each s_i we can find $r_i \in \mathbb{Z}$, such that

$$m r_i - \frac{m}{2} < s_i < m r_i + \frac{m}{2}$$

(strict inequality holds because m is odd). Write $s_i = m r_i + t_i$, with $|t_i| < \frac{m}{2}$. Set $\gamma = r_0 + r_1 i + r_2 j + r_3 k$; then $N(\sigma - \gamma m) = t_0^2 + t_1^2 + t_2^2 + t_3^2 < 4 \left(\frac{m}{2}\right)^2 = m^2$; this proves the claim.

To prove the lemma, set $m = N(\beta) = \beta \overline{\beta}$ and $\sigma = \alpha \overline{\beta}$. By the claim we can find $\gamma \in \mathbb{H}(\mathbb{Z})$, such that

$$N(\beta) N(\overline{\beta}) = N(\beta)^2 = m^2 > N(\sigma - \gamma m) = N(\alpha \overline{\beta} - \gamma \beta \overline{\beta})$$
$$= N(\alpha - \gamma\beta) N(\overline{\beta}).$$

Setting $\delta = \alpha - \gamma\beta$ and cancelling out $N(\overline{\beta})$, we get $N(\delta) < N(\beta)$, as required. $\square$

Note that the left-hand Euclidean algorithm provides for $\gamma_1, \delta_1 \in \mathbb{H}(\mathbb{Z})$, such that $\alpha = \beta \gamma_1 + \delta_1$, with $N(\delta_1) < N(\beta)$.

2.6.4. Definition. Let α, β be integral quaternions. We say that $\delta \in \mathbb{H}(\mathbb{Z})$ is a *right-hand greatest common divisor* of α and β if

(a) δ is a right-hand divisor of α and β;
(b) if $\delta_0 \in \mathbb{H}(\mathbb{Z})$ is a right-hand divisor of both α and β, then δ_0 is a right-hand divisor of δ.

We denote such a δ by $(\alpha, \beta)_r$; it is clear that $(\alpha, \beta)_r$ is unique up to associate, if it exists. We plan to show that, under suitable conditions, $(\alpha, \beta)_r$ indeed exists.

2.6.5. Lemma. Let $\alpha \in \mathbb{H}(\mathbb{Z})$. Then α has a unique factorization:

$$\alpha = 2^\ell \, \pi \, \alpha_0,$$

where $\ell \in \mathbb{N}, \pi \in \{1, 1+i, 1+j, 1+k, (1+i)(1+j), (1+i)(1-k)\}$ and $\alpha_0 \in \mathbb{H}(\mathbb{Z})$ is odd.

Proof. We indicate how to prove existence, leaving the proof of uniqueness as an exercise. So fix $\alpha \in \mathbb{H}(\mathbb{Z})$; let 2^ℓ be the highest power of 2 dividing α; set $\alpha' = \frac{\alpha}{2^\ell}$, and write

$$\alpha' = a_0 + a_1 i + a_2 j + a_3 k,$$

where at least one of the a_i's is odd. Since multiplication by a unit changes the position of the a_i's, up to associate we may assume that a_0 is odd. Now, if α' is odd, then $\alpha = 2^\ell \alpha'$ and we are finished. Therefore, we may assume α' even, and two cases then occur.

(a) $N(\alpha') \equiv 2 \pmod{4}$.

Then exactly two a_i's are odd, with a_0 among them. If, say, a_0 and a_1 are odd, then

$$\alpha_0 = \frac{a_0 + a_1}{2} + \left(\frac{a_1 - a_0}{2} \right) i + \left(\frac{a_2 + a_3}{2} \right) j + \left(\frac{a_3 - a_2}{2} \right) k$$

is in $\mathbb{H}(\mathbb{Z})$, it is odd, and $\alpha' = (1+i)\alpha_0$. The other possibilities (a_0 and a_2 odd, or a_0 and a_3 odd) allow one to factor out either $1+j$ or $1+k$ and are treated in the same way:

(b) $N(\alpha') \equiv 0 \pmod{4}$.

Then all the a_i's are odd, therefore congruent to ± 1 (mod. 4). In any case $N(\alpha') \equiv 4$ (mod. 8). In this case we need to consider the possibilities for various combinations of congruences modulo 4, in principle sixteen different subcases. However, one finds that these can be grouped into two families of eight subcases each, depending on whether an even or an odd number of the a_i's are congruent to 1 (mod. 4).

Claim A. If an even number of the a_i's are congruent to 1 (mod. 4), then there exists an odd quaternion α_1, such that $\alpha' = (1 + i)(1 + j)\alpha_1$.

Proof. First note that, multiplying by a unit if necessary, we may assume that $a_0 \equiv 1$ (mod. 4). First assume that $a_0 \equiv a_1 \equiv 1$ (mod. 4) and $a_2 \equiv a_3 \equiv \pm 1$ (mod. 4). As in case (a), we then have $\alpha' = (1 + i)\alpha_0$ with

$$\alpha_0 = \frac{a_0 + a_1}{2} + \left(\frac{a_0 - a_1}{2}\right) i + \left(\frac{a_2 + a_3}{2}\right) j + \left(\frac{a_3 - a_2}{2}\right) k.$$

Notice that $\frac{a_0 + a_1}{2}$ and $\frac{a_2 + a_3}{2}$ are odd, while $\frac{a_0 - a_1}{2}$ and $\frac{a_3 - a_2}{2}$ are even. By case (a), we then have $\alpha_0 = (1 + j)\alpha_1$, where α_1 is odd since $N(\alpha_0) \equiv 2$ (mod. 4). Hence, $\alpha' = (1 + i)(1 + j)\alpha_1$ as desired.

Assume now that $a_0 \equiv a_2 \equiv 1$ (mod. 4) and $a_1 \equiv a_3 \equiv \pm 1$ (mod. 4). Proceeding as before, we may write $\alpha' = (1 + j)(1 + k)\alpha_1$, with α_1 odd. Notice then that $(1 + j)(1 + k) = (1 + i)(1 + j)$. The last case, $a_0 \equiv a_3 \equiv 1$ (mod. 4) and $a_1 \equiv a_2 \equiv \pm 1$ (mod. 4), is entirely similar, using $(1 + k)(1 + i) = (1 + i)(1 + j)$.

Claim B. If an odd number of the a_i's are congruent to 1 (mod. 4), then there exists an odd quaternion α_1, such that $\alpha' = (1 + i)(1 - k)\alpha_1$.

Proof. Again we may assume without loss of generality that three of the a_i's are congruent to 1 (mod. 4), with a_0 among them. If $a_0 \equiv a_1 \equiv a_2 \equiv 1$ (mod. 4) and $a_3 \equiv -1$ (mod. 4), then as in case (a) we have $\alpha' = (1 + i)\alpha_0$ with

$$\alpha_0 = \frac{a_0 + a_1}{2} + \left(\frac{a_0 - a_1}{2}\right) i + \left(\frac{a_2 + a_3}{2}\right) j + \left(\frac{a_3 - a_2}{2}\right) k$$
$$= b_0 + b_1 i + b_2 j + b_3 k.$$

Now b_0 and b_3 are odd, while b_1 and b_2 are even. Then

$$\alpha_0 = (1 - k)\left(\frac{b_0 + b_3}{2} + \left(\frac{b_1 - b_2}{2}\right) i + \left(\frac{b_1 + b_2}{2}\right) j + \left(\frac{b_0 + b_3}{2}\right) k\right)$$
$$= (1 - k)\alpha_1,$$

where α_1 is odd. So $\alpha_0 = (1+i)(1-k)\alpha_1$. The remaining cases are shown in an analogous way. $\square$

We let $\mathbb{Z}\left[\frac{1}{2}\right]$ denote the subring of rational numbers defined by

$$\mathbb{Z}\left[\frac{1}{2}\right] = \left\{\frac{k}{2^n} : k \in \mathbb{Z}, n \in \mathbb{N}\right\}.$$

2.6.6. Theorem. Let $\alpha, \beta \in \mathbb{H}(\mathbb{Z})$, with β odd. Then $(\alpha, \beta)_r$ exists. Moreover, the following version of Bézout's relation holds: there exist $\gamma, \delta \in \mathbb{H}\left(\mathbb{Z}\left[\frac{1}{2}\right]\right)$ such that $(\alpha, \beta)_r = \gamma\,\alpha + \delta\,\beta$.

Proof. We mimic the proof of the Euclidean algorithm for the greatest common divisor of two integers. By Lemma 2.6.3, we find $\gamma_0, \delta_0 \in \mathbb{H}(\mathbb{Z})$, with $N(\delta_0) < N(\beta)$, such that

$$\alpha = \gamma_0\,\beta + \delta_0.$$

By Lemma 2.6.5, we have $\delta_0 = 2^{\ell_0}\,\pi_0\,\delta_0'$, with δ_0' odd and $N(\delta_0') \le N(\delta_0) < N(\beta)$. Again by Lemmas 2.6.3 and 2.6.5,

$$\beta = \gamma_1\,\delta_0' + \delta_1,$$

and $\delta_1 = 2^{\ell_1}\,\pi_1\,\delta_1'$, with $N(\delta_1') \le N(\delta_1) < N(\delta_0')$ and δ_1' odd. By repeated applications of Lemmas 2.6.3 and 2.6.5, we find quaternions $\gamma_i, \delta_i, \delta_i' \in \mathbb{H}(\mathbb{Z})$, such that

$$\delta_{i-1}' = \gamma_{i+1}\,\delta_i' + \delta_{i+1},$$

and $\delta_{i+1} = 2^{\ell_{i+1}}\,\pi_{i+1}\,\delta_{i+1}'$, with $N(\delta_{i+1}') \le N(\delta_{i+1}) < N(\delta_i')$, and δ_{i+1}' odd. The last two equations are

$$\delta_{k-2}' = \gamma_k\,\delta_{k-1}' + \delta_k$$
$$\delta_{k-1}' = \gamma_{k+1}\,\delta_k',$$

since the δ_i's are a sequence of quaternions in $\mathbb{H}(\mathbb{Z})$ with strictly decreasing norms. We claim that $(\alpha, \beta)_r = \delta_k'$. Indeed, δ_k' is clearly a right-hand divisor of $\delta_{k-1}', \delta_{k-2}', \ldots, \delta_1', \beta, \alpha$. And if δ is a right-hand divisor of α and β, then it is a right-hand divisor of δ_0, hence also of δ_0', by the uniqueness part of Lemma 2.6.5; going on, we see that δ is a right-hand divisor of δ_k'.

Finally, we rewrite the previous system as

$$\delta_0' = 2^{-\ell_0} \pi_0^{-1}(\alpha - \gamma_0\,\beta)$$
$$\delta_1' = 2^{-\ell_1} \pi_1^{-1}(\beta - \gamma_1\,\delta_0')$$
$$\vdots$$
$$\delta_k' = 2^{-\ell_k} \pi_k^{-1}(\delta_{k-2}' - \gamma_k\,\delta_{k-1}').$$

Since π_i is invertible in $\mathbb{H}\left(\mathbb{Z}\left[\frac{1}{2}\right]\right)$, this expresses δ_k' as

$$\delta_k' = \gamma\,\alpha + \delta\,\beta,$$

with $\gamma, \delta \in \mathbb{H}\left(\mathbb{Z}\left[\frac{1}{2}\right]\right)$. $\quad\square$

Following the line of argument employed for $\mathbb{Z}$ and $\mathbb{Z}\,[i]$, we now wish to use the greatest common right-hand divisor and the modified Bézout relation to characterize the primes of $\mathbb{H}(\mathbb{Z})$ and to develop a theory of factorization in this ring. Along the way, we will see that every rational prime has a nontrivial factorization into two conjugate prime quaternions. Ultimately we will show that α is a prime quaternion in $\mathbb{H}(\mathbb{Z})$ if and only if $N(\alpha)$ is a rational prime.

As in $\mathbb{Z}\,[i]$, divisibility properties of quaternions are related to divisibility properties of their norms. The following lemma is analogous to Lemma 2.2.9.

2.6.7. Lemma. For $\alpha \in \mathbb{H}(\mathbb{Z})$ and $m \in \mathbb{Z}$, m odd,

$$(m, \alpha)_r = 1 \quad \text{if and only if} \quad (m, N(\alpha))_r = 1.$$

Proof. We first prove the direct implication, so assume that $(m, \alpha)_r = 1$. By Bézout's relation 2.6.6, there exists $\gamma, \delta \in \mathbb{H}\left(\mathbb{Z}\left[\frac{1}{2}\right]\right)$ with

$$(m, \alpha)_r = 1 = \gamma\,m + \delta\,\alpha.$$

Then,

$$N(\delta)\,N(\alpha) = N(1 - \gamma m) = (1 - \gamma m)(1 - \overline{\gamma}m)$$
$$= 1 - (\gamma + \overline{\gamma})m + N(\gamma)m^2$$

or

$$1 = N(\delta)\,N(\alpha) + (\gamma + \overline{\gamma})m - N(\gamma)m^2.$$

Since $N(\delta)$, $N(\gamma)$ and $\gamma + \overline{\gamma}$ are elements of $\mathbb{Z}\left[\frac{1}{2}\right]$, we can find $k \in \mathbb{N}$, such that $2^k\,N(\delta)$, $2^k(\gamma + \overline{\gamma})$, $2^k\,N(\gamma)$ are rational integers. Let $\beta \in \mathbb{H}(\mathbb{Z})$ be a right-hand divisor of both $N(\alpha)$ and m; since m is odd, β is an odd quaternion.

From

$$2^k = (2^k N(\delta)) N(\alpha) + (2^k (\gamma + \overline{\gamma})) m - (2^k N(\gamma)) m^2,$$

we see that β is a right-hand divisor of 2^k. Taking norms, we see that $N(\beta)$ divides 2^{2k}. Since $N(\beta)$ is odd, we must have $N(\beta) = 1$; in other words, β is invertible.

The proof of the converse is completely similar to the proof of the corresponding statement in Lemma 2.2.9. □

2.6.8. Lemma. Let $p \in \mathbb{N}$ be an odd, rational prime. Assume that there exists $\alpha \in \mathbb{H}(\mathbb{Z})$, such that α is not divisible by p, but that $N(\alpha)$ is divisible by p. Set $(\alpha, p)_r = \delta$. Then δ is prime in $\mathbb{H}(\mathbb{Z})$, and $N(\delta) = p$.

Proof. Write $p = \gamma \delta$, for some $\gamma \in \mathbb{H}(\mathbb{Z})$. First we notice that γ is not a unit; otherwise, p would be associate to δ and hence would divide α, contradicting our assumption. Next, since p divides $N(\alpha)$, it follows from Lemma 2.6.7 that δ is not a unit. On the other hand, taking norms we get

$$p^2 = N(p) = N(\gamma) N(\delta),$$

with $N(\gamma) \neq 1 \neq N(\delta)$. So we must have $N(\gamma) = N(\delta) = p$.

From $N(\delta) = p$, it follows that δ is prime in $\mathbb{H}(\mathbb{Z})$. Indeed if $\delta = xy$ is a factorization of δ in $\mathbb{H}(\mathbb{Z})$, taking norms we get $N(\delta) = p = N(x) N(y)$, so either $N(x) = 1$ or $N(y) = 1$; in either case x or y is a unit. □

2.6.9. Theorem. For every odd, rational prime $p \in \mathbb{N}$, there exists a prime $\delta \in \mathbb{H}(\mathbb{Z})$, such that $N(\delta) = p = \delta \overline{\delta}$. In particular, p is not prime in $\mathbb{H}(\mathbb{Z})$.

Proof. By Proposition 2.5.3, there exist $x, y \in \mathbb{Z}$, such that $1 + x^2 + y^2 \equiv 0$ (mod. p). Set $\alpha = 1 + xi + yj$; clearly p does not divide α, but p divides $N(\alpha) = 1 + x^2 + y^2$. So Lemma 2.6.8 applies, and $\delta = (\alpha, p)_r$ is the desired prime in $\mathbb{H}(\mathbb{Z})$. □

Finally we are able to show the following:

2.6.10. Corollary. $\delta \in \mathbb{H}(\mathbb{Z})$ is prime in $\mathbb{H}(\mathbb{Z})$ if and only if $N(\delta)$ is prime in $\mathbb{Z}$.

Proof. We have seen in the course of the proof of Lemma 2.6.8 that if $N(\delta)$ is a rational prime then δ is prime in $\mathbb{H}(\mathbb{Z})$. So only the direct implication needs to be proven. Let δ be a prime in $\mathbb{H}(\mathbb{Z})$.

Assume first that δ is even. By Lemma 2.6.5, we have $\delta = 2^\ell \pi \delta_0$, where $\ell \in \mathbb{N}$, $\pi \in \{1, 1+i, 1+j, 1+k, (1+i)(1+j), (1+i)(1-k)\}$, and δ_0 is odd. Note that 2 is not prime in $\mathbb{H}(\mathbb{Z})$ since $2 = (1+i)(1-i)$. Since by assumption δ is prime in $\mathbb{H}(\mathbb{Z})$, we must have $\ell = 0$, $N(\delta_0) = 1$ (since δ_0 is odd), and $\pi \in \{1+i, 1+j, 1+k\}$, so that $N(\delta) = 2$, as required.

Now suppose that δ is odd. Let $p \in \mathbb{N}$ be an odd, rational prime dividing $N(\delta)$. We must show that $N(\delta) = p$. Set $\alpha = (p, \delta)_r$; then $\delta = \gamma \alpha$ for some $\gamma \in \mathbb{H}(\mathbb{Z})$. It follows from Lemma 2.6.7 that α is not a unit in $\mathbb{H}(\mathbb{Z})$. Since δ is prime in $\mathbb{H}(\mathbb{Z})$, we see that γ must be a unit in $\mathbb{H}(\mathbb{Z})$, so that α and δ are associate. Hence, δ is a right-hand divisor of p, say $p = \psi \delta$ for some $\psi \in \mathbb{H}(\mathbb{Z})$. Taking norms and remembering that p divides $N(\delta)$, we get

$$p = N(\psi) \left(\frac{N(\delta)}{p} \right).$$

If $N(\psi) = 1$, then p and δ are associate, so that p is prime in $\mathbb{H}(\mathbb{Z})$, which contradicts Theorem 2.6.9. Hence, $\frac{N(\delta)}{p} = 1$, so $N(\delta) = p$. $\square$

As a consequence of the arithmetic of $\mathbb{H}(\mathbb{Z})$, we get Lagrange's celebrated result on sums of four squares, which of course also follows from Theorem 2.4.1.

2.6.11. Corollary. Every natural number is a sum of four squares.

Proof. Let $n \in \mathbb{N}$. The result is obvious for $n = 0$ and $n = 1$, so we may assume $n \geq 2$. Let $n = 2^{r_0} p_1^{r_1} \dots p_k^{r_k}$ be the factorization of n into primes, with p_i odd. By Theorem 2.6.9, we can find $\delta_i \in \mathbb{H}(\mathbb{Z})$, such that $p_i = N(\delta_i) = \delta_i \overline{\delta}_i$, while $2 = (1+i)(1-i)$. Hence, each prime appearing in n can be written as a sum of squares, and the multiplicativity of the quaternionic norm gives the final representation of n itself in that form. $\square$

As the example following Proposition 2.6.2 shows, we cannot expect unique factorization into primes in $\mathbb{H}(\mathbb{Z})$. We will now restrict attention to the set of integral quaternions α with $N(\alpha) = p^k$, where p is an odd, rational prime. We will see that, for these α, we can recover a sort of unique factorization for these α's.

So let p be an odd prime. By Jacobi's Theorem 2.4.1,

$$a_0^2 + a_1^2 + a_2^2 + a_3^2 = p$$

has $8(p+1)$ integral solutions, each corresponding to an integral quaternion $\alpha = a_0 + a_1 i + a_2 j + a_3 k$ of norm p. If $p \equiv 1 \pmod 4$, one a_i is odd, while the rest are even; if $p \equiv 3$ (mod. 4), one a_i is even, while the rest are odd. In each case, one coordinate, call it a_i^0, is distinguished. If $a_i^0 \neq 0$, then among the eight associates $\varepsilon \alpha$ of α; exactly one will have $|a_i^0|$ as its zero-th component. (Note the absolute value here!). If $a_i^0 = 0$, as it might when $p \equiv 3$ (mod. 4), then two associates, $\varepsilon \alpha$ and $-\varepsilon \alpha$, will each have $a_0 = 0$. In this case we may designate either one as distinguished.

Hence, there are $p + 1$ distinguished solutions of

$$a_0^2 + a_1^2 + a_2^2 + a_3^2 = p,$$

such that the corresponding quaternion α satisfies either $\alpha \equiv 1$ (mod. 2) or $\alpha \equiv i + j + k$ (mod. 2). In this list of solutions, both α and $\overline{\alpha}$ appear whenever $a_0 > 0$, while only one of the pair is included when $a_0 = 0$. We thus form the set

$$S_p = \{\alpha_1, \overline{\alpha}_1, \ldots, \alpha_s, \overline{\alpha}_s, \beta_1, \ldots, \beta_t\},$$

where α_i has $a_0^{(i)} > 0$, β_j has $b_0^{(j)} = 0$, and $\alpha_i \overline{\alpha}_i = -\beta_j^2 = p$. Note that $2s + t = |S_p| = p + 1$.

2.6.12. Definition. A *reduced word* over S_p is a word over the alphabet S_p, which has no subword of the form $\alpha_i \overline{\alpha}_i, \overline{\alpha}_i \alpha_i, \beta_j^2$ $(i = 1, \ldots, s; j = 1, \ldots, t)$. The *length* of a word is the number of occurring symbols.

2.6.13. Theorem. Let $k \in \mathbb{N}$; let $\alpha \in \mathbb{H}(\mathbb{Z})$ be such that $N(\alpha) = p^k$. Then α admits a unique factorization $\alpha = \varepsilon \, p^r \, w_m$, where ε is a unit in $\mathbb{H}(\mathbb{Z})$, w_m is a reduced word of length m over S_p, and $k = 2r + m$.

Proof. We begin with existence. So fix $\alpha \in \mathbb{H}(\mathbb{Z})$ with $N(\alpha) = p^k$. By Proposition 2.6.2, α is a product of primes in $\mathbb{H}(\mathbb{Z})$:

$$\alpha = \delta_1 \ldots \delta_n.$$

By Corollary 2.6.10, we must have $N(\delta_i) = p$ $(1 \leq i \leq n)$, and therefore $n = k$. Since $N(\delta_i) = p$, we find a unit ε_i and $\gamma_i \in S_p$, such that $\delta_i = \varepsilon_i \gamma_i$; hence,

$$\alpha = \varepsilon_1 \gamma_1 \varepsilon_2 \gamma_2 \ldots \varepsilon_k \gamma_k.$$

Now, it is easy to see that for every $\gamma \in S_p$ and every unit ε of $\mathbb{H}(\mathbb{Z})$, we can find some $\gamma' \in S_p$ and some unit ε', such that $\gamma \varepsilon = \varepsilon' \gamma'$. In the previous factorization of α, this allows all the ε_i's to be moved to the left and to write

$$\alpha = \varepsilon \, \gamma_1' \ldots \gamma_k',$$

with $\gamma_i' \in S_p$ and ε a unit in $\mathbb{H}(\mathbb{Z})$. So we have written α as the product of a unit and a word over the alphabet S_p, but this word is not necessarily reduced. We make it reduced simply by moving any factor p to the left, if there is an occurrence of $\alpha_i \overline{\alpha}_i$, $\overline{\alpha}_i \alpha_i$ or β_j^2 in the word: we then get a shorter word, for which we iterate the process: this proves existence.

We prove uniqueness by a counting argument. First, by Jacobi's theorem 2.4.1, there are exactly

$$8 \sum_{i=0}^{k} p^i = 8 \left(\frac{p^{k+1} - 1}{p - 1} \right)$$

quaternions $\alpha \in \mathbb{H}(\mathbb{Z})$ with $N(\alpha) = p^k$. Now we count the number of reduced words of length m over the alphabet S_p: there are $p + 1$ possible choices for the first letter, and p possible choices for each of the following letters (since we have to avoid subwords of the form $\alpha_i \overline{\alpha}_i$, $\overline{\alpha}_i \alpha_i$ and β_j^2). Thus, the number of reduced words of length m is

$$\begin{cases} 1 & \text{if} \quad m = 0 \\ (p+1)\, p^{m-1} & \text{if} \quad m \geq 1. \end{cases}$$

Hence, the total number of expressions of the form $\varepsilon\, p^r\, w_m$, with ε a unit, w_m a reduced word of length m, and $2r + m = k$, is

$$\begin{cases} 8 \left(1 + \displaystyle\sum_{r=0}^{\frac{k}{2}-1} (p+1)\, p^{k-2r-1} \right) & \text{if } k \text{ is even,} \\[6pt] 8 \displaystyle\sum_{r=0}^{\frac{k-1}{2}} (p+1)\, p^{k-2r-1} & \text{if } k \text{ is odd.} \end{cases}$$

In both cases, we find $8 \left(\frac{p^{k+1}-1}{p-1} \right)$ expressions, which coincide with the number of $\alpha \in \mathbb{H}(\mathbb{Z})$ with $N(\alpha) = p^k$. Since, by the existence part, every such α can be written in such a form, this factorization must be unique. $\square$

We denote by Λ' the following subset of $\mathbb{H}(\mathbb{Z})$:

$$\Lambda' = \{\alpha = a_0 + a_1 i + a_2 j + a_3 k \in \mathbb{H}(\mathbb{Z}) : \alpha \equiv 1 \ (\mathrm{mod.}\ 2)$$

or

$$\alpha \equiv i + j + k \ (\mathrm{mod.}\ 2),\ N(\alpha) \text{ a power of } p \}.$$

It is easy to see, by reducing mod. 2, that Λ' is closed under multiplication; clearly it contains S_p.

2.6.14. Corollary. Every element $\alpha \in \Lambda'$ with $N(\alpha) = p^k$ has a unique factorization $\alpha = \pm\, p^r\, w_m$, where $r \in \mathbb{N}$, w_m is a reduced word of length m over S_p, and $k = 2r + m$.

Proof. By Theorem 2.6.13, α can be written in a unique way as $\alpha = \varepsilon\, p^r\, w_m$, with r and w_m having the desired properties and ε as a unit in $\mathbb{H}(\mathbb{Z})$. Reducing mod. 2, we get $\alpha \equiv \varepsilon\, w_m$ (mod. 2). Any $\alpha_i, \beta_j \in S_p$ that appears in w_m has $\alpha_i, \beta_j \equiv 1$ (mod. 2) or $\alpha_i, \beta_j \equiv i + j + k$ (mod. 2). For the moment, denote the latter as γ. Then, in modulo 2, we have the congruences:

$$\alpha \equiv \begin{cases} \varepsilon & \text{if an even number of } \gamma\text{'s appears in } w_m; \\[2mm] \varepsilon(i + j + k) & \text{if an odd number of } \gamma\text{'s appears in } w_m. \end{cases}$$

On the other hand, since $\alpha \in \Lambda'$, α itself must satisfy $\alpha \equiv 1$ (mod. 2) or $\alpha \equiv i + j + k$ (mod. 2). Therefore, we see that in every case we must have $\varepsilon \equiv 1$ (mod. 2); in other words, $\varepsilon = \pm 1$. $\square$

Exercises on Section 2.6

1. Prove the uniqueness part in Lemma 2.6.5.

2. For $\gamma \in S_p$ and ε a unit in $\mathbb{H}(\mathbb{Z})$, show that there exist $\gamma' \in S_p$ and a unit ε' in $\mathbb{H}(\mathbb{Z})$, such that $\gamma\, \varepsilon = \varepsilon'\, \gamma'$.

3. Look at the example following Proposition 2.6.2 (nonuniqueness of factorizations of 13 in $\mathbb{H}(\mathbb{Z})$); how do you reconciliate it with Theorem 2.6.13 (uniqueness of factorization)?

2.7. Notes on Chapter 2

2.2. A good introduction to sums of two, three, and four squares can be found in Landau [37]. The proof that (i) $\Leftrightarrow$ (ii) in the Fermat – Euler theorem 2.1.7 is taken from [1]: it is a "proof from The Book," in the sense of Erdös. Legendre's formula 2.1.11 was first proved by Jacobi, using the theory of elliptic functions. The proof we give is taken from Hardy and Wright [32; 16.10]. There is no such simple formula for $r_3(n)$ as there is for $r_2(n)$ or $r_4(n)$. However, there is a criterion of Gauss: $r_3(n) > 0$ (or, n is a sum of three squares) if and only if n is *not* of the form $4^a(8b - 1)$, with $a, b \in \mathbb{N}$ (see [61], Appendix of Chapter 4, this work). From this result, it is a simple exercise to deduce Lagrange's theorem that $r_4(n) > 0$ for every n and, hence, that every positive integer is a sum of four squares.

2.3. Our proof of quadratic reciprocity 2.3.2 is taken from Hardy and Wright, [32; 6.13].

2.4. The elementary proof of Jacobi's formula 2.4.1 given in section 2.4 is due to Dirichlet [21] and was communicated by him in a letter to Liouville. This proof was advocated by Weil in [67]. We recall that Jacobi also computed $r_4(n)$ for arbitrary n: the general formula is

$$r_4(n) = 8 \sum_{\substack{d|n \\ d \not\equiv 0 \, (\mathrm{mod.}\, 4)}} d.$$

This has also been proved by means of the theory of elliptic functions; see [32; 20.11] for a proof.

2.6. The factorization theory of odd integral quaternions is due to Dickson [19]. A more complete theory for slightly more general integral quaternions was obtained by Hurwitz [35] (see also [56]). The reader can also refer to the discussion of Hurwitz quaternions in Hardy and Wright [32].

Chapter 3
$\mathrm{PSL}_2(q)$

3.1. Some Finite Groups

The Ramanujan graphs $X^{p,q}$, to be defined in Chapter 4, will be associated with the finite groups $\mathrm{PGL}_2(q)$ and $\mathrm{PSL}_2(q)$ that we define in this section.

Let K be a field. We denote by $\mathrm{GL}_2(K)$ the group of invertible 2-by-2 matrices with coefficients in K, that is, those matrices with nonzero determinant. We denote by $\mathrm{SL}_2(K)$ the subgroup of matrices with determinant 1, which forms the kernel of the determinant map,

$$\det : \mathrm{GL}_2(K) \to K^{\times}.$$

We denote by $\mathrm{PGL}_2(K)$ the quotient group

$$\mathrm{PGL}_2(K) = \mathrm{GL}_2(K) \Big/ \left\{ \begin{pmatrix} \lambda & 0 \\ 0 & \lambda \end{pmatrix} : \lambda \in K^{\times} \right\},$$

and by $\mathrm{PSL}_2(K)$ the quotient group

$$\mathrm{PSL}_2(K) = \mathrm{SL}_2(K) \Big/ \left\{ \begin{pmatrix} \varepsilon & 0 \\ 0 & \varepsilon \end{pmatrix} : \varepsilon = \pm 1 \right\}.$$

The two latter groups can be realized more concretely as follows. Let $P^1(K) = K \cup \{\infty\}$ be the projective line over K. We will embed $\mathrm{PGL}_2(K)$ and $\mathrm{PSL}_2(K)$ into the group $\mathrm{Sym}\, P^1(K)$ of permutations of $P^1(K)$. To every $A = \begin{pmatrix} a & b \\ c & d \end{pmatrix} \in \mathrm{GL}_2(K)$, we associate the fractional linear transformation (or Möbius transformation),

$$\varphi_A : P^1(K) \to P^1(K),$$

defined by

$$\varphi_A(z) = \frac{az + b}{cz + d}.$$

Here we set $\varphi_A(\infty) = \begin{cases} \frac{a}{c} & \text{if } c \neq 0 \\ \infty & \text{if } c = 0 \end{cases}$ and $\varphi_A\left(\frac{-d}{c}\right) = \infty$. Thus, we get a group homomorphism

$$\varphi : GL_2(K) \to \operatorname{Sym} P^1(K),$$

where

$$\varphi(A) = \varphi_A,$$

and $PGL_2(K)$ (resp. $PSL_2(K)$) identifies with $\varphi(GL_2(K))$ (resp. $\varphi(SL_2(K))$).

When $K = \mathbb{F}_q$, the finite field of order q, we write $GL_2(q)$, $SL_2(q)$, $PGL_2(q)$, and $PSL_2(q)$ for the four groups defined previously.

3.1.1. Proposition.

(a) $|GL_2(q)| = q(q-1)(q^2-1)$

(b) $|SL_2(q)| = |PGL_2(q)| = q(q^2-1)$

(c) $|PSL_2(q)| = \begin{cases} q(q^2-1) & \text{if } q \text{ is even} \\ \frac{q(q^2-1)}{2} & \text{if } q \text{ is odd.} \end{cases}$

Proof. (a) A 2-by-2 matrix in $GL_2(q)$ is obtained by first choosing the first column, a nonzero vector in $\mathbb{F}_q^2$: there are $q^2 - 1$ possible choices for that; then by choosing the second column, a vector in $\mathbb{F}_q^2$ linearly independent from the first one: there are $q^2 - q$ possible choices for that.

(b) and (c) follow from elementary group theory. $\square$

Exercises on Section 3.1

1. (a) For $A, B \in GL_2(K)$, prove that $\varphi_{AB} = \varphi_A \circ \varphi_B$.

   (b) Check carefully that $\operatorname{Ker} \varphi$ is exactly the subgroup $\left\{ \begin{pmatrix} \lambda & 0 \\ 0 & \lambda \end{pmatrix} : \lambda \in K^\times \right\}$ of scalar matrices.

2. (a) Give details of the proof of Proposition 3.1.1, (b) (c).

   (b) True or false: is $SL_2(q)$ isomorphic to $PGL_2(q)$?

3. For $A \in GL_2(K)$, show that $\varphi_A \in PSL_2(K)$ if and only if $\det A$ is a square in $K^\times$.

3.2. Simplicity

The properties of the Ramanujan graphs $X^{p,q}$ in Chapter 4 will depend on some structural properties of $PSL_2(q)$. Simplicity will be used, on one hand, to determine which $X^{p,q}$'s are bipartite and, on the other hand, to establish the expanding properties of the $X^{p,q}$'s.

3.2.1. Lemma. For any field K, the group SL$_2$(K) is generated by the union of the two subgroups $\left\{ \begin{pmatrix} 1 & \lambda \\ 0 & 1 \end{pmatrix} : \lambda \in K \right\}$ and $\left\{ \begin{pmatrix} 1 & 0 \\ \mu & 1 \end{pmatrix} : \mu \in K \right\}$. Hence, every matrix in SL$_2$(K) is a finite product of matrices which are either upper-triangular or lower-triangular and which have 1's along the diagonal.

Proof. Let $\begin{pmatrix} a & b \\ c & d \end{pmatrix} \in$ SL$_2$(K). We distinguish two cases:

(a) $c \neq 0$. Then an immediate computation gives

$$\begin{pmatrix} 1 & \frac{a-1}{c} \\ 0 & 1 \end{pmatrix} \begin{pmatrix} 1 & 0 \\ c & 1 \end{pmatrix} \begin{pmatrix} 1 & \frac{d-1}{c} \\ 0 & 1 \end{pmatrix} = \begin{pmatrix} a & \frac{a(d-1)}{c} + \frac{a-1}{c} \\ c & d \end{pmatrix}.$$

However, $\frac{a(d-1)}{c} + \frac{a-1}{c} = \frac{ad-1}{c} = \frac{ad-(ad-bc)}{c} = b$.

(b) $c = 0$. Then $d \neq 0$, and the matrix $\begin{pmatrix} a+b & b \\ d & d \end{pmatrix} \in$ SL$_2$(K) can be treated as in the first case. But then

$$\begin{pmatrix} a+b & b \\ d & d \end{pmatrix} \begin{pmatrix} 1 & 0 \\ -1 & 1 \end{pmatrix} = \begin{pmatrix} a & b \\ 0 & d \end{pmatrix}. \qquad \square$$

Recall that a group G is *simple* if its only normal subgroups are $\{1\}$ and G; equivalently, every group homomorphism $\pi : G \to H$ is either constant or one-to-one. The following result was proved by Jordan in 1861.

3.2.2. Theorem. Let K be a field with $|K| \geq 4$. Then PSL$_2$(K) is a simple group.

Proof. Using the homomorphism $\varphi : $ SL$_2$(K) $\to$ PSL$_2$(K), it is enough to show that a normal subgroup N of SL$_2$(K), not contained in Ker $\varphi = \left\{ \begin{pmatrix} \varepsilon & 0 \\ 0 & \varepsilon \end{pmatrix} : \varepsilon = \pm 1 \right\}$, is equal to SL$_2$($K$). So let A be a nonscalar matrix in N. Since A is nonscalar, there exists a vector $v \in K^2$ which is not an eigenvector of A, so v and Av are linearly independent over K. This means that $\{v, Av\}$ is a basis of K^2. (Note the crucial role of dimension 2 in this proof). In that basis, A is written $\begin{pmatrix} 0 & b \\ 1 & d \end{pmatrix}$, and since det $A = 1$, we actually have $A = \begin{pmatrix} 0 & -1 \\ 1 & d \end{pmatrix}$. We now appeal to a classical trick: for every $B \in N$, $C \in$ SL$_2$(K), the commutator $C^{-1} B^{-1} C B$ is in N (since $C^{-1} B^{-1} C \in N$ and $B \in N$). We first apply this trick with $B = A$,

$$C = \begin{pmatrix} \alpha & 0 \\ 0 & \alpha^{-1} \end{pmatrix} \ (\alpha \in K^{\times}); \text{ then}$$

$$C^{-1} A^{-1} C A = \begin{pmatrix} \alpha^{-2} & d(\alpha^{-2} - 1) \\ 0 & \alpha^2 \end{pmatrix} \in N.$$

Repeating the trick with $B' = \begin{pmatrix} \alpha^{-2} & d(\alpha^{-2} - 1) \\ 0 & \alpha^2 \end{pmatrix}$ and $C' = \begin{pmatrix} 1 & \mu \\ 0 & 1 \end{pmatrix}$ ($\mu \in K$), we get

$$C'^{-1} B'^{-1} C' B' = \begin{pmatrix} 1 & \mu(\alpha^4 - 1) \\ 0 & 1 \end{pmatrix} \in N.$$

If $|K| \geq 4$ and $|K| \neq 5$, we can find $\alpha \in K^{\times}$, such that $\alpha^4 \neq 1$. Then the set of values of $\mu(\alpha^4 - 1)$ ($\mu \in K$, $\alpha \in K^{\times}$) is exactly K, so that $N \supseteq \left\{ \begin{pmatrix} 1 & \lambda \\ 0 & 1 \end{pmatrix} : \lambda \in K \right\}$. Since

$$\begin{pmatrix} 0 & -1 \\ 1 & 0 \end{pmatrix} \begin{pmatrix} 1 & -\mu \\ 0 & 1 \end{pmatrix} \begin{pmatrix} 0 & -1 \\ 1 & 0 \end{pmatrix}^{-1} = \begin{pmatrix} 1 & 0 \\ \mu & 1 \end{pmatrix},$$

one also has $N \supseteq \left\{ \begin{pmatrix} 1 & 0 \\ \mu & 1 \end{pmatrix} : \mu \in K \right\}$. By Lemma 3.2.1, we have $N = \mathrm{SL}_2(K)$.

This concludes the proof for $K \neq \mathbb{F}_5$; only the remaining case makes the proof lengthy. Although we will not need it, for completeness we give the proof for $K = \mathbb{F}_5$ as well.

So assume $K = \mathbb{F}_5$. From the first part of the previous proof, we have $\begin{pmatrix} \alpha^{-2} & d(\alpha^{-2} - 1) \\ 0 & \alpha^2 \end{pmatrix} \in N$, for every $\alpha \in \mathbb{F}_5^{\times}$. Take $\alpha = 2$; then $\begin{pmatrix} -1 & -2d \\ 0 & -1 \end{pmatrix} \in N$, and, squaring, we see that $\begin{pmatrix} 1 & -d \\ 0 & 1 \end{pmatrix} \in N$. Two cases are possible:

(a) $d \neq 0$. The powers of $\begin{pmatrix} 1 & -d \\ 0 & 1 \end{pmatrix}$ are then $\left\{ \begin{pmatrix} 1 & \lambda \\ 0 & 1 \end{pmatrix} : \lambda \in \mathbb{F}_5 \right\}$, and we conclude as for general fields that $N = \mathrm{SL}_2(\mathbb{F}_5)$.

(b) $d = 0$, so $A = \begin{pmatrix} 0 & -1 \\ 1 & 0 \end{pmatrix}$. We then perform the standard trick with $B = A$ and $C'' = \begin{pmatrix} \delta & 1 \\ -1 & 0 \end{pmatrix}$ ($\delta \in \mathbb{F}_5^{\times}$), so that

$$A' = C''^{-1} A^{-1} C'' A = \begin{pmatrix} 1 & -\delta \\ -\delta & \delta^2 + 1 \end{pmatrix} \in N.$$

Since A' is nonscalar, in a suitable basis of $\mathbb{F}_5^2$ it will take the form $A' = \begin{pmatrix} 0 & -1 \\ 1 & d' \end{pmatrix}$, as at the beginning of the proof, with

$$d' = \operatorname{Tr} A' = \delta^2 + 2.$$

Now the squares in $\mathbb{F}_5^{\times}$ are ± 1, so that either $d' = 1$ or $d' = 3$. In any case $d' \neq 0$, so case (a) applies to A', and the proof is complete. $\square$

Exercises on Section 3.2

(Note: Alt(n) denotes the alternating group on n letters.)

1. Show that PSL$_2(2)$ is isomorphic to Sym(3) and also to the group of isometries of an equilateral triangle; deduce that it is not a simple group.

2. Show that PSL$_2(3)$ is isomorphic to Alt(4) and also to the group of rotations of a regular tetrahedron; deduce that it is not a simple group.

3. Show that PSL$_2(4)$ and PSL$_2(5)$ are isomorphic to Alt(5).

3.3. Structure of Subgroups

To establish the property of connectedness for the Ramanujan graphs $X^{p,q}$ to be constructed in Chapter 4, we will need to understand some of the structure of the subgroups of PSL$_2(q)$.

We will make repeated use of the following general principle: Recall that, if σ is a permutation of a set X, and $x \in X$, the *orbit of x under σ* is $\Omega_x = \{\sigma^k(x) : k \in \mathbb{Z}\}$.

3.3.1. Lemma. Let σ be a permutation of a set X. If σ has prime order p, then every orbit of σ on X has either 1 or p elements.

Proof. Let H be the subgroup generated by σ in Sym(X). For $x \in X$, it is a general fact that $|\Omega_x| = \frac{|H|}{|H_x|}$, where $H_x = \{\alpha \in H : \alpha(x) = x\}$ is the stabilizer of x in H. Here, $|H| = p$ by assumption, so that either $|H_x| = 1$ and $|\Omega_x| = p$ or $|H_x| = p$ and $|\Omega_x| = 1$. (Note that if $|\Omega_x| = 1$, then x is a fixed point of σ.) $\square$

As a first application of this principle, we prove Cauchy's theorem on the existence of elements with prime orders in finite groups. Eventually, this result of Cauchy was superseded by Sylow's theorems, but since we will not need the later, stronger results, we prove just the earlier one here.

3.3.2. Theorem. Let G be a finite group, and let p be a prime. If p divides $|G|$, then G contains some element of order p.

Proof. Consider the product $G^p = G \times \cdots \times G$ (p factors). Let σ be the cyclic permutation of factors:

$$\sigma(g_1, g_2, \ldots, g_p) = (g_2, \ldots, g_p, g_1).$$

Clearly σ is a permutation of G^p, with order p. Now let H be the subset of G^p defined by

$$H = \{(g_1, \ldots, g_p) \in G^p : g_1 g_2 \ldots g_p = 1\}.$$

Clearly $|H| = |G|^{p-1}$, since we may freely choose the $p - 1$ first coordinates of a p-tuple in H. If $g_1 g_2 \ldots g_p = 1$, then conjugating by g_1^{-1} we get $g_2 \ldots g_p g_1 = 1$, meaning that H is invariant under σ. From now on, we view σ as a permutation of H. Since orbits of σ partition H, and since $|H|$ is a multiple of p by assumption, we see that the sum of orders of orbits of σ in H is congruent to 0 modulo p. By Lemma 3.3.1, orbits of σ are either fixed points or have p elements. Since σ has at least one fixed point in H, namely, the p-tuple $(1, 1, \ldots, 1)$, it must have at least $p - 1$ other ones to match the previous congruence. Such a fixed point is clearly of the form $(g, g, \ldots, g)$, with $g \neq 1$. To say that this p-tuple is in H means exactly that $g^p = 1$; i.e., g has order p in G. This concludes the proof. $\square$

Now we need a group-theoretical definition.

3.3.3. Definition. A group G is *metabelian* if it admits a normal subgroup N such that both N and G/N are abelian.

In particular, abelian groups are metabelian and metabelian groups are solvable; subgroups of metabelian groups are metabelian.

In 1901, Dickson gave a list, up to isomorphism, of all subgroups of $PSL_2(q)$, where q is a prime power. We refer to [34], Theorem 8.27, for the precise statement. Specializing to the case where q is a prime, and looking up Dickson's list, one notices that all proper subgroups of $PSL_2(q)$ are metabelian, with two possible exceptions:

- Sym(4), of order 24, which is solvable but not metabelian;
- Alt(5), of order 60, which is simple nonabelian.

Our purpose in this section is to give a direct proof of this fact.

3.3.4. Theorem. Let q be a prime. Let H be a proper subgroup of $PSL_2(q)$, such that $|H| > 60$. Then H is metabelian.

Theorem 3.3.4 immediately follows from the following two results.

3.3.5. Proposition. Let q be a prime, and let H be a proper subgroup of $PSL_2(q)$. If q divides $|H|$, then H is metabelian.

3.3.6. Proposition. Let q be a prime, and let H be a subgroup of $PSL_2(q)$. If $|H| > 60$ and q does not divide $|H|$, then H has an abelian subgroup of index at most 2; in particular, H is metabelian. (Note that by Proposition 3.1.1, if q does not divide $|H|$, then H is a proper subgroup.)

To prove Proposition 3.3.5, we first need a description of elements of order q in $PSL_2(q)$. Recall that $\varphi : SL_2(q) \to PSL_2(q)$ defined by $\varphi(A) = \varphi_A$ denotes the canonical map. For simplicity of notation, let C_b denote the matrix

$$C_b = \begin{pmatrix} 1 & b \\ 0 & 1 \end{pmatrix}.$$

3.3.7. Lemma. Let q be a prime. For $A \in SL_2(q)$, the following properties are equivalent:

(i) φ_A has order q;
(ii) there is a unique one-dimensional subspace D in $\mathbb{F}_q^2$ such that either A or $-A$ fixes D pointwise;
(iii) φ_A is conjugate in $PGL_2(q)$ to some φ_{C_b}, with $b \in \mathbb{F}_q^\times$.

Proof. (i) $\Rightarrow$ (ii) We recall that φ_A is a fractional linear transformation on $P^1(\mathbb{F}_q)$. Since $|P^1(\mathbb{F}_q)| = 1 + q$ and φ_A has order q, it follows from Lemma 3.3.1 that φ_A has a unique fixed point on $P^1(\mathbb{F}_q)$. The latter corresponds to a unique one-dimensional subspace D in $\mathbb{F}_q^2$ which is globally invariant under A. Now A has order q or $2q$ in $SL_2(q)$, and we examine both cases:

(a) A has order q. Since A acts on D with at least one fixed point (namely, $(0, 0)$) and $|D| = q$, it follows from Proposition 3.1.1 that A fixes D pointwise.

(b) A has order $2q$. Then the preceding argument applies to A^2, so A^2 fixes D pointwise. Then A acts on D by $x \mapsto -x$ so that $-A$ fixes D pointwise.

(ii) $\Rightarrow$ (iii) Choose a basis $\{e_1, e_2\}$ of $\mathbb{F}_q^2$, with $e_1 \in D$. The matrix of A in this basis has the form $\begin{pmatrix} a & b \\ 0 & d \end{pmatrix}$, with $a = d = \pm 1$ and $b \neq 0$. This means that, in $\mathrm{PGL}_2(q)$, the transformation φ_A is conjugate to $\varphi_{C_{ab}}$.

(iii) $\Rightarrow$ (i) This is immediate, since φ_{C_b} has order q. $\quad\square$

The previous proof actually shows the following: let $A, B \in \mathrm{SL}_2(q)$ be such that φ_A and φ_B have order q; if A, B globally fix the same line D in $\mathbb{F}_q^2$, then φ_A and φ_B generate the same subgroup of order q in $\mathrm{PSL}_2(q)$.

Proof of Proposition 3.3.5. Since q divides $|H|$, it follows from Theorem 3.3.2 that H contains at least one subgroup of order q.

Claim. H contains a unique subgroup of order q. Indeed, suppose by contradiction that C_1, C_2 are distinct subgroups of order q in H. By Lemma 3.3.7 and the remark following it, they correspond to two distinct lines D_1, D_2 in $\mathbb{F}_q^2$. Choose a basis $\{e_1, e_2\}$ of $\mathbb{F}_q^2$, with $e_i \in D_i$ ($i = 1, 2$). Working in this basis, we have $\quad C_1 = \varphi \left\{ \begin{pmatrix} 1 & \lambda \\ 0 & 1 \end{pmatrix} : \lambda \in \mathbb{F}_q \right\}$ and $C_2 = \varphi \left\{ \begin{pmatrix} 1 & 0 \\ \mu & 1 \end{pmatrix} : \mu \in \mathbb{F}_q \right\}$. By Lemma 3.2.1, the subgroup generated by the union of C_1 and C_2 is $\mathrm{PSL}_2(q)$; this contradicts the assumption that H is a proper subgroup of $\mathrm{PSL}_2(q)$.

So let C be the unique subgroup of order q in H. By uniqueness, C is normal in H. Conjugating if necessary within $\mathrm{PGL}_2(q)$, we may assume, by Lemma 3.3.7, that

$$C = \varphi \left\{ \begin{pmatrix} 1 & \lambda \\ 0 & 1 \end{pmatrix} : \lambda \in \mathbb{F}_q \right\},$$

so that the action of C on the projective line $P^1(\mathbb{F}_q)$ is by translations, $z \mapsto z + \lambda$. Since the unique fixed point of C in $P^1(\mathbb{F}_q)$ is ∞, and C is normal in H, we have, for every $\varphi_A \in C$, $\varphi_B \in H$,

$$\varphi_A(\varphi_B(\infty)) = \varphi_B(\varphi_{B^{-1}AB}(\infty)) = \varphi_B(\infty),$$

so that $\varphi_B(\infty)$ is fixed under C. Thus, $\varphi_B(\infty) = \infty$ for every $\varphi_B \in H$, which means that H is contained in the stabilizer of ∞ in $\mathrm{PSL}_2(q)$. But this is nothing but the subgroup

$$B_0 = \varphi \left\{ \begin{pmatrix} a & b \\ 0 & a^{-1} \end{pmatrix} : a \in \mathbb{F}_q^\times, \, b \in \mathbb{F}_q \right\},$$

sometimes called the standard Borel subgroup of $\text{PSL}_2(q)$. Since B_0 is meta-belian, so is H. $\square$

Before proving Proposition 3.3.6, we need some more terminology.

3.3.8. Definition. Let G be a group; let $J \subseteq H \subseteq G$ be subgroups; let $g \in H$.

(a) The *centralizer* $C_H(g)$ of g in H is the subgroup of elements in H that commute with g:

$$C_H(g) = \{h \in H : hg = gh\}.$$

(b) The *normalizer* $N_H(J)$ of J in H is the subgroup of elements in H that normalize J:

$$N_H(J) = \{h \in H : h J h^{-1} = J\}.$$

3.3.9. Lemma. Let G be a finite group, and let Z be a central subgroup of G. Assume that, for every $g \in G - Z$, the centralizer $C_G(g)$ is abelian. Let J, K be maximal abelian subgroups of G. If $J \neq K$, then $J \cap K = Z$.

Proof. We first notice that every maximal abelian subgroup J of G must contain Z. Indeed, since Z is central, $JZ = ZJ$ is an abelian subgroup containing J. By maximality, we must have $JZ = J$; i.e., $J \supseteq Z$.

Claim. For every $g \in G - Z$, the centralizer $C_G(g)$ is a maximal abelian subgroup of G. Indeed, let J be a maximal abelian subgroup containing $C_G(g)$. Since J commutes with g, we must have $J \subseteq C_G(g)$; i.e., $J = C_G(g)$.

The lemma is now easily proved. If J, K are maximal abelian subgroups in G with $J \cap K \neq Z$, we find $g \in (J \cap K) - Z$. Then $C_G(g)$ is maximal abelian, and $J \subseteq C_G(g)$, $K \subseteq C_G(g)$ since J, K are abelian. By maximality we must have $J = C_G(g) = K$. $\square$

Notice that the assumption in Lemma 3.3.9 is inherited by any subgroup of G containing Z. We now show that this assumption is satisfied by $\text{SL}_2(q)$, q prime.

3.3.10. Lemma. Let q be a prime. Every nonscalar matrix in $\text{SL}_2(q)$ has an abelian centralizer.

Proof. Let $A = \begin{pmatrix} a & b \\ c & d \end{pmatrix}$ be a nonscalar matrix in $SL_2(q)$. We consider the fractional linear transformation φ_A on $P^1(\mathbb{F}_{q^2})$, the projective line over the field with q^2 elements. Since A is nonscalar, we have $\varphi_A \neq \mathrm{Id}$. The fixed-point equation

$$\frac{az + b}{cz + d} = z$$

has one or two solutions in $P^1(\mathbb{F}_{q^2})$. We separate cases:

(a) φ_A has a unique fixed point: Conjugating within $PGL_2(q^2)$, we may assume that this fixed point is ∞; then φ_A is a translation:

$$\varphi_A(z) = z + b \qquad (z \in \mathbb{F}_{q^2}),$$

so $A = \pm \begin{pmatrix} 1 & b \\ 0 & 1 \end{pmatrix}$. The centralizer of A in $SL_2(q^2)$ is the subgroup $\left\{ \pm \begin{pmatrix} 1 & \lambda \\ 0 & 1 \end{pmatrix}, \ \lambda \in \mathbb{F}_{q^2} \right\}$, which is abelian.

(b) φ_A has two fixed points; conjugating within $PGL_2(q^2)$, we may assume that these are 0 and ∞. Then $\varphi_A(z) = \alpha^2 z$ for some $\alpha \in \mathbb{F}_{q^2}^\times, \alpha \neq \pm 1$. This means $A = \pm \begin{pmatrix} \alpha & 0 \\ 0 & \alpha^{-1} \end{pmatrix}$.

The centralizer of A inside $SL_2(q^2)$ is then the diagonal subgroup, which is abelian. $\square$

3.3.11. Lemma. Let q be an odd prime. Let H be a subgroup of $SL_2(q)$, containing scalar matrices, with q not dividing $|H|$. If J is a maximal abelian subgroup of H, then $[N_H(J) : J] \leq 2$.

Proof. The result is obvious when H is the subgroup of scalar matrices in $SL_2(q)$. So we may assume that H, and hence also J, contains some nonscalar matrix A. As in the proof of Lemma 3.3.10, we consider the fractional linear transformation φ_A on $P^1(\mathbb{F}_{q^2})$.

Claim. φ_A has two fixed points on $P^1(\mathbb{F}_{q^2})$. Indeed if φ_A has only one fixed point, then A is conjugate in $L = SL_2(q^2)$ to $\pm \begin{pmatrix} 1 & b \\ 0 & 1 \end{pmatrix}$; hence, A has order q or $2q$. So q divides $|H|$, which is a contradiction.

Conjugating within L, we may assume that the fixed points of φ_A are $\{0, \infty\}$. Since J is abelian, we have $J \subseteq C_L(A)$, where

$$C_L(A) = \left\{ \begin{pmatrix} a & 0 \\ 0 & a^{-1} \end{pmatrix} : a \in \mathbb{F}_{q^2}^{\times} \right\}.$$

Now, for $g \in J$, $n \in N_H(J)$ and $z \in \{0, \infty\}$, we have

$$g(n(z)) = n(n^{-1} g n(z)) = n(z),$$

since $n^{-1} g n \in J$. So $n(z)$ is fixed under J, so that $n(z) \in \{0, \infty\}$. This shows that every element in $N_H(J)$ acts as a permutation on $\{0, \infty\}$, and this defines a homomorphism

$$\omega : N_H(J) \to \mathrm{Sym}\,\{0, \infty\}.$$

Clearly the kernel Ker ω is contained in $C_L(A)$, and Ker ω is therefore abelian. On the other hand, Ker ω contains J. Since J is a maximal abelian subgroup of H, we must have Ker $\omega = J$. So $[N_H(J) : J] = |N_H(J)/J| \leq |\mathrm{Sym}\,\{0, \infty\}| = 2$, and the proof is concluded. $\square$

Proof of Proposition 3.3.6. Let H be a subgroup of PSL$_2(q)$, with $|H| > 60$, and q not dividing $|H|$. In view of Proposition 3.1.1, we must have $q \geq 7$; in particular, q is odd. So

$$\varphi : \mathrm{SL}_2(q) \to \mathrm{PSL}_2(q)$$

has a kernel of order 2. Set $\widetilde{H} = \varphi^{-1}(H)$, and $|\widetilde{H}| = 2h$. We denote by $C_1, \ldots, C_s$ the conjugacy classes of maximal abelian subgroups J of $\widetilde{H}$ with $[N_{\widetilde{H}}(J) : J] = 1$ and by $C_{s+1}, \ldots, C_{s+t}$ the conjugacy classes of maximal abelian subgroups J of $\widetilde{H}$ with $[N_{\widetilde{H}}(J) : J] = 2$. By Lemma 3.3.11, these are the only possibilities. Note that, since $\widetilde{H}$ contains at least one maximal abelian subgroup, we have $s + t \geq 1$. For J_i, a representative in C_i, set $|J_i| = 2 g_i$.

Claim. For every nonscalar matrix $A \in \widetilde{H}$, there exists a unique index i ($1 \leq i \leq s + t$), such that A is conjugate within $\widetilde{H}$ to some element of J_i.

Existence is clear since A is contained in some maximal abelian subgroup of $\widetilde{H}$, itself conjugate to some J_i.

For uniqueness, let us assume that A is conjugate to some element of J_i and to some element of J_j:

$$B_i A B_i^{-1} \in J_i \quad \text{and} \quad B_j A B_j^{-1} \in J_j,$$

for some $B_i, B_j \in \tilde{H}$. Then $A \in B_i^{-1} J_i B_i \cap B_j^{-1} J_j B_j$. By Lemma 3.3.10, the group $\tilde{H}$ satisfies the assumption in Lemma 3.3.9. Since $B_i^{-1} J_i B_i$ and $B_j^{-1} J_j B_j$ are maximal abelian subgroups of $\tilde{H}$, it follows from Lemma 3.3.9 that $B_i^{-1} J_i B_i = B_j^{-1} J_j B_j$; J_i and J_j are conjugate within $\tilde{H}$, so $i = j$. This proves the claim.

From the claim, it follows that, for fixed i, the number of nonscalar matrices in $\tilde{H}$, which are conjugate to some element of J_i, is $(|J_i| - 2) \cdot |C_i|$. But $|C_i| = \frac{|\tilde{H}|}{|N_{\tilde{H}}(J_i)|} = \frac{|\tilde{H}|}{|J_i| [N_{\tilde{H}}(J_i):J_i]}$ so $(|J_i| - 2) |C_i| = \frac{(g_i - 1) 2h}{g_i [N_{\tilde{H}}(J_i):J_i]}$; therefore,

$$2h - 2 = \sum_{i=1}^{s} \frac{(g_i - 1) 2h}{g_i} + \sum_{j=s+1}^{s+t} \frac{(g_j - 1) 2h}{2 g_j};$$

this leads to the *basic relation*:

$$1 = \frac{1}{h} + \sum_{i=1}^{s} \left(1 - \frac{1}{g_i}\right) + \sum_{j=s+1}^{s+t} \frac{1}{2}\left(1 - \frac{1}{g_j}\right).$$

Now $g_i, g_j \geq 2$; hence, $1 - \frac{1}{g_i} \geq \frac{1}{2}$ and

$$1 \geq \frac{1}{h} + \frac{s}{2} + \frac{t}{4} > \frac{s}{2} + \frac{t}{4}.$$

The inequality $1 > \frac{s}{2} + \frac{t}{4}$ has exactly five integral solutions with $s \geq 0, t \geq 0$, and $s + t \geq 1$:

(a) $s = 1, t = 0$;
(b) $s = 1, t = 1$;
(c) $s = 0, t = 1$;
(d) $s = 0, t = 2$;
(e) $s = 0, t = 3$.

We now examine these solutions case by case.

(a) The basic relation gives $1 = \frac{1}{h} + 1 - \frac{1}{g_1}$, i.e., $h = g_1$: then $\tilde{H} = J_1$; i.e., $\tilde{H}$ is abelian, so that H is abelian.

(b) The basic relation becomes $1 = \frac{1}{h} + 1 - \frac{1}{g_1} + \frac{1}{2}\left(1 - \frac{1}{g_2}\right)$, or $\frac{1}{g_1} + \frac{1}{2g_2} = \frac{1}{2} + \frac{1}{h}$. Now $\frac{1}{g_1} + \frac{1}{4} \geq \frac{1}{g_1} + \frac{1}{2g_2} > \frac{1}{2}$, so that $2 \leq g_1 < 4$.

Claim. $g_1 = 2$. If not, then $g_1 = 3$, leading to $\frac{1}{3} + \frac{1}{2g_2} > \frac{1}{2}$; i.e., $g_2 < 3$, or $g_2 = 2$. Then, from the basic relation, we get $h = 12$, contradicting $h > 60$.

From $g_1 = 2$, we deduce $h = 2g_2$; i.e., $[\tilde{H} : J_2] = 2$ and $[H : \varphi(J_2)] = 2$: H has an abelian subgroup with index 2.

(c) Actually this case is impossible: indeed the basic relation gives $1 = \frac{1}{h} + \frac{1}{2} - \frac{1}{2g_1}$, i.e. $\frac{1}{2} + \frac{1}{2g_1} = \frac{1}{h}$. This contradicts the inequality $|\tilde{H}| = 2h \geq |N_{\tilde{H}}(J_1)| = 4g_1$.

(d) Also this case is impossible. Indeed the basic relation gives $1 = \frac{1}{h} + \frac{1}{2} - \frac{1}{2g_1} + \frac{1}{2} - \frac{1}{2g_2}$, or $\frac{1}{h} = \frac{1}{2}\left(\frac{1}{g_1} + \frac{1}{g_2}\right)$.

By Lemma 3.3.9, the subgroup $J_1 \cap J_2$ is exactly the subgroup of scalar matrices; i.e., $|J_1 \cap J_2| = 2$. So $2h = |\tilde{H}| \geq |J_1 J_2| = 2g_1 g_2$. Hence, $\frac{1}{h} = \frac{1}{2}\left(\frac{g_1 + g_2}{g_1 g_2}\right) \geq \frac{1}{2}\frac{g_1 + g_2}{h}$; i.e., $g_1 + g_2 \leq 2$, which contradicts $g_1 \geq 2, g_2 \geq 2$.

(e) The basic relation becomes $1 = \frac{1}{h} + \frac{1}{2} - \frac{1}{2g_1} + \frac{1}{2} - \frac{1}{2g_2} + \frac{1}{2} - \frac{1}{2g_3}$, which gives $\frac{1}{2g_1} + \frac{1}{2g_2} + \frac{1}{2g_3} = \frac{1}{h} + \frac{1}{2} > \frac{1}{2}$. Clearly we may assume $g_1 \leq g_2 \leq g_3$.

- We first notice that $g_1 = 2$. Indeed, assuming $g_1 \geq 3$, we get $\frac{1}{2g_1} + \frac{1}{2g_2} + \frac{1}{2g_3} \leq \frac{1}{2}$, which contradicts the previous inequality. Then $\frac{1}{2g_2} + \frac{1}{2g_3} = \frac{1}{h} + \frac{1}{4} > \frac{1}{4}$.
- We now observe that $g_2 = 2$. Indeed, if we had $g_2 \geq 4$, we would get $\frac{1}{2g_2} + \frac{1}{2g_3} \leq \frac{1}{4}$, which contradicts the previous inequality. But if $g_2 = 3$, then $\frac{1}{2g_3} = \frac{1}{h} + \frac{1}{12}$, where

$$\frac{1}{12} < \frac{1}{2g_3} = \frac{1}{h} + \frac{1}{12} < \frac{1}{60} + \frac{1}{12} = \frac{1}{10};$$

i.e., $6 > g_3 > 5$, which cannot happen.
- So we get, from the basic relation: $h = 2g_3$; i.e., $[\tilde{H} : J_3] = 2$ and $[H : \varphi(J_3)] = 2$. As in case (b), H has an abelian subgroup with index 2. $\square$

Exercises on Section 3.3

1. Check the details of the following implication: if a group H has an abelian subgroup of index 2, then H is metabelian.

2. Let G be a group. For $g_1, g_2 \in G$, define the commutator of g_1, g_2 as $[g_1, g_2] = g_1 g_2 g_1^{-1} g_2^{-1}$. Show that G is metabelian if and only if, for every $g_1, g_2, g_3, g_4 \in G$,

$$[[g_1, g_2], [g_3, g_4]] = 1.$$

3. Let K be a field. Show that the following groups are metabelian.
   (a) the affine group of K, i.e., the group of permutations of the form $z \mapsto az + b$ $(a \in K^\times, b \in K)$.
   (b) the three-dimensional Heisenberg group over K,

$$H_3(K) = \left\{ \begin{pmatrix} 1 & x & z \\ 0 & 1 & y \\ 0 & 0 & 1 \end{pmatrix} : x, y, z \in K \right\}.$$

4. Let H be a subgroup of $PSL_2(q)$, where q is a prime. Assume that $|H| > 60$ and that q does not divide $|H|$. Show that H is either cyclic or dihedral [Hint: in the proof of Proposition 3.3.6, show that cases (a), (b), and (e) correspond, respectively, to subgroups which are cyclic, dihedral of order $2n$ (n odd), and dihedral of order $2n$ (n even).]

5. The purpose of this exercise is to see that Lemmas 3.3.10 and 3.3.11 become false upon replacing $SL_2(q)$ by $PSL_2(q)$.
   (a) Consider the fractional linear transformation $z \mapsto -z$ of $P^1(\mathbb{F}_q)$. Show that, if $q \equiv 1$ (mod. 4), it belongs to $PSL_2(q)$. Compute then its centralizer in $PSL_2(q)$.
   (b) For a suitable value of q, construct a subgroup H of $PSL_2(q)$ and a maximal abelian subgroup J of H such that $[N_H(J) : J] = 3$ [Hint: use exercise 3 in section 3.2.]

6. Let p be a prime; set $q = p^r$, with $r \geq 2$. Show that Theorem 3.3.4 does not hold for $PSL_2(q)$. [Hint: remember that $\mathbb{F}_p$ is a subfield of $\mathbb{F}_q$.]

3.4. Representation Theory of Finite Groups

Students frequently ask: "We worked so hard in studying some group theory, why should we learn group representations on top of that?"

The answer is twofold: first, in a certain sense, a group is a nonlinear object, and linear representations are a way to linearize it. Second, linear algebra is a powerful tool that sheds more light on groups themselves.[1]

A representation of a group G on a real or complex vector space V is a homomorphism from G into the linear group of V, that is, into the group of invertible linear transformations on V. If V is finite-dimensional and a basis of V has been fixed, this means that a representation of G is a homomorphism into the group of matrices with nonzero determinant acting on V.

---

[1] Philosophically, this is analogous to what happens in calculus, where a certain affine space, namely, the tangent space at a point of the graph of a function, provides valuable information about the underlying differentiable function in a neighborhood of that point.

Both in physics and in mathematics, groups appear most naturally as symmetries of some object X. (Recall, for example, that Sym (3), the symmetric group on three letters, can be realized as the group of symmetries of an equilateral triangle; similarly, D_4, the dihedral group of order 8, can be realized as the symmetries of a square.) A convenient rephrasing of the fact that G is a symmetry group of X says that X is a G-space; in other words, we are given a homomorphism from G to the group of permutations of X. Then G also acts on any object associated with X and, in particular, on functions on X. Since functions on X form a vector space $\mathbb{C}X$, we get a representation λ_X of G on $\mathbb{C}X$, where each element $g \in G$ acts on $\mathbb{C}X$ by taking the function $f \in \mathbb{C}X$ to the function $\lambda_X(g) f \in \mathbb{C}X$, defined by

$$(\lambda_X(g) f)(x) = f(g^{-1} x) \qquad (x \in X).$$

Suppose we can find a linear operator T on $\mathbb{C}X$, which commutes with the representation λ_X; i.e.,

$$\lambda_X(g) T = T \lambda_X(g)$$

for every $g \in G$. Suppose that the function f is an eigenfunction of T, associated with the eigenvalue μ; then so is the function $\lambda_X(g) f$, since

$$T \lambda_X(g) f = \lambda_X(g) T f = \mu \lambda_X(g) f.$$

So the eigenspace V_μ of T corresponding to μ is a subspace of $\mathbb{C}X$ which is *invariant* under $\lambda_X(G)$; in other words, eigenspaces of commuting operators allow us to decompose the representation into smaller pieces. (The theme of decomposing representations into subspaces of smaller dimension is a recurrent one; representations which cannot be decomposed further are said to be *irreducible*: these are the building blocks of the theory.)

Specifically, suppose that $X = (V, E)$ is a finite graph. Let G be a group of automorphisms of X, and let λ_V be the corresponding representation of G on $\mathbb{C}V$. Since G maps edges to edges, λ_X commutes with the adjacency matrix A. Since A is diagonalizable, we can use eigenspaces of A to start decomposing λ_V into irreducibles; or conversely, we can use *a priori* knowledge on the representations of G to bound multiplicities of eigenvalues of A. In section 3.5, we shall prove that any nontrivial representation of $\mathrm{PSL}_2(q)$ has dimension at least $\frac{q-1}{2}$; in Chapter 4 we shall use that information to see that, for the graphs $X^{p,q}$ constructed there, the multiplicity of the nontrivial eigenvalues is at least $\frac{q-1}{2}$.

3.4.1. Definition. Let G be a group. A *representation* of G is a pair (π, V), where V is a complex vector space and π is a homomorphism $G \to GL(V)$. (Here $GL(V)$ is the group of linear permutations of V.) The *degree* of (π, V) is the dimension $\dim_\mathbb{C} V$ of V.

When there is no risk of confusion about which vector space is involved, we write π instead of (π, V).

3.4.2. Examples.

(i) The constant homomorphism $G \to GL(V)$ defines the trivial representation of G on V.

(ii) Every homomorphism $G \to \mathbb{C}^\times$ gives rises to a representation of degree 1 of G on $\mathbb{C}$.

(iii) Let X be a G-space, that is, a nonempty set endowed with a homomorphism $G \to \mathrm{Sym}(X)$, where $\mathrm{Sym}(X)$ is the group of permutations of X. Let $\mathbb{C}X$ be the set of functions $X \to \mathbb{C}$ that are zero except on finite subsets of X. The *permutation representation* λ_X of G on $\mathbb{C}X$ is defined as earlier by

$$(\lambda_X(g)f)(x) = f(g^{-1} \cdot x),$$

where $f \in \mathbb{C}X$, $g \in G$, $x \in X$.

(iv) Viewing G as a G-space by means of left multiplication, we get the *left regular representation* λ_G of G on $\mathbb{C}G$:

$$(\lambda_G(g)f)(x) = f(g^{-1}x) \qquad (f \in \mathbb{C}G \,;\, g, x \in G).$$

Viewing G as a G-space by means of right multiplication, we get the *right regular representation* ρ_G of G on $\mathbb{C}G$:

$$(\rho_G(g)f)(x) = f(xg) \qquad (f \in \mathbb{C}G \,;\, g, x \in G).$$

To analyze a representation, it makes it easier to have invariant subspaces.

3.4.3. Definition. Let (π, V) be a representation of G. A linear subspace W of V is *invariant* if for every $g \in G : \pi(g)(W) = W$.

If W is invariant, then $(\pi \mid_W, W)$ is also a representation of G, called a *subrepresentation* of π. The subspaces 0 and V are the *trivial* invariant subspaces.

3.4.4. Example. Let X be a G-space. Set

$$W_0 = \left\{ f \in \mathbb{C}X : \sum_{x \in X} f(x) = 0 \right\} ;$$

this is an invariant subspace of λ_X. If X is finite, then W_0 admits a G-invariant complement: the subspace W_1 of constant functions on X. Note that $\lambda_X \mid_{W_1}$ is the trivial representation of degree 1 on W_1.

3.4.5. Definition. A representation (π, V) with $V \neq \{0\}$ is *irreducible* if it has no nontrivial invariant subspace.

3.4.6. Examples.
   (i) Every representation of degree 1 is irreducible.
   (ii) The canonical representation of $GL(V)$ on V (given by the identity homomorphism $GL(V) \to GL(V)$) is irreducible, since $GL(V)$ acts transitively on the set of linear subspaces of the same dimension.

As in most of mathematics, the notion of equivalence plays a crucial role in representation theory. Here we determine equivalence of representations of a group G through the existence of certain linear maps on the associated representation spaces.

3.4.7(a). Definition. Let (π, V) and (ρ, W) be two representations of G. A linear map $T : V \to W$ *intertwines* π *and* ρ if, for every $g \in G$, one has $T\,\pi\,(g) = \rho\,(g)\,T$. We denote by $\mathrm{Hom}_G(\pi, \rho)$ the vector space of intertwiners between π and ρ.

3.4.7(b). Definition. Let (π, V) and (ρ, W) be two representations of G. We say that π and ρ are *equivalent* if there exists an invertible intertwiner in $\mathrm{Hom}_G(\pi, \rho)$, that is, a linear map $T : V \to W$, such that, for every $g \in G$,

$$\rho\,(g) = T\,\pi\,(g)\,T^{-1}.$$

Note that Definition 3.4.7(a) means that whenever T intertwines π and ρ, then the following diagram commutes:

$$
\begin{array}{ccc}
V & \xrightarrow{\ \pi(g)\ } & V \\
{\scriptstyle T}\downarrow & & \downarrow{\scriptstyle T} \\
W & \xrightarrow{\ \rho(g)\ } & W
\end{array}
$$

3.4.8. Examples.
   (i) Consider the map $T : \mathbb{C}G \to \mathbb{C}G$ defined by

$$(Tf)(x) = f(x^{-1}) \qquad (f \in \mathbb{C}G,\ x \in G).$$

Then $T \in \mathrm{Hom}_G(\lambda_G, \rho_G)$; since $T^2 = \mathrm{Id}$, we see, furthermore, that λ_G and ρ_G are equivalent.

(ii) For a G-space X, consider the map $T : \mathbb{C}X \to \mathbb{C}$ given by

$$f \mapsto \sum_{x \in X} f(x);$$

T intertwines λ_X and the trivial representation of degree 1.

The next result is the celebrated Schur's lemma; its proof sheds light on the role of complex vector spaces in representation theory.

3.4.9. Theorem. Let (π, V), (ρ, W) be finite-dimensional, irreducible representations of G. Then

$$\dim_{\mathbb{C}} \mathrm{Hom}_G(\pi, \rho) = \begin{cases} 0 & \text{if } \pi \text{ and } \rho \text{ are not equivalent;} \\ 1 & \text{if } \pi \text{ and } \rho \text{ are equivalent.} \end{cases}$$

Proof. We prove the first equality by contraposition, so assume that $\dim_{\mathbb{C}}$ $\mathrm{Hom}_G(\pi, \rho) > 0$. Then there exists a nonzero intertwiner T from π to ρ. We must show that π and ρ are equivalent. First, the kernel, $\mathrm{Ker}\, T$, is an invariant subspace of π with $\mathrm{Ker}\, T \neq V$ by assumption. Since π is irreducible, we must have $\mathrm{Ker}\, T = \{0\}$, so T is injective. Next, the image, $\mathrm{Im}\, T$, is a nonzero invariant subspace of ρ. Since ρ is irreducible, we have $\mathrm{Im}\, T = W$, so T is onto. Thus, T is invertible, and we have shown that π and ρ are equivalent.

To prove the second equality, since π and ρ are equivalent, we may assume that $\pi = \rho$. Now $\mathrm{Hom}_G(\pi, \pi)$ always contains the one-dimensional subspace of scalar matrices αI; hence, it will be enough to show that any intertwiner $T \in \mathrm{Hom}_G(\pi, \pi)$ is scalar when π is irreducible. As V is a finite-dimensional *complex* vector space, the linear operator T on V has at least one eigenvalue $\lambda \in \mathbb{C}$; that is,

$$\mathrm{Ker}\,(T - \lambda \cdot \mathrm{Id}_V) \neq 0.$$

Since $\mathrm{Ker}\,(T - \lambda \cdot \mathrm{Id}_V)$ is an invariant subspace and π is irreducible, we must have $\mathrm{Ker}\,(T - \lambda \cdot \mathrm{Id}_V) = V$; i.e., $T = \lambda \cdot \mathrm{Id}_V$. $\square$

An alternative statement of Schur's Lemma reads as follows: Let (π, V) and (ρ, W) be two irreducible representations of G, and let T intertwine π and ρ. Then either (a) $f \equiv 0$ or (b) f is an isomorphism; thus, π is equivalent to ρ and $V \cong W$. In this case f is a scalar map $f(v) = \lambda v$ for some $\lambda \in \mathbb{C}$.

From now on, we shall consider representations of finite groups on finite-dimensional complex vector spaces only.

We first define the tensor product $V \otimes W$ of two finite-dimensional complex vector spaces V and W.

We identify $V \otimes W$ with a vector space of formal products in the following way. Consider the Cartesian product $V \times W$ of pairs (v, w) with $v \in V$, $w \in W$, and form the additive group consisting of finite sums of pairs with complex coefficients; that is,

$$G = \left\{ \sum \alpha_{ij}(v_i, w_j) \mid \alpha_{ij} \in \mathbb{C}, v_i \in V, w_j \in W \right\}.$$

Let H be the subgroup of G generated by the subset of sums of the form

  (i) $(v_1 + v_2, w) - (v_1, w) - (v_2, w)$,
  (ii) $(v, w_1 + w_2) - (v, w_1) - (v, w_2)$,
  (iii) $(v, \alpha w) - (\alpha v, w)$,

where $\alpha \in \mathbb{C}$. Now define a map $i : V \times W \to G/H$ by setting

$$i(v, w) = (v, w) + H.$$

The group G/H constructed previously, which itself forms a vector space over $\mathbb{C}$, is called the *tensor product* of V and W; it is denoted by $V \otimes W$.

We can be even more specific in describing elements of $V \otimes W$. If $\{v_i\}_{1 \le i \le n}$ and $\{w_j\}_{1 \le j \le m}$ are sets of basis vectors for V and W, respectively, then the set $\{i(v_i, w_j)\}$ is a basis for $V \otimes W$. Informally, $V \otimes W$ can be thought of as a set of finite sums of products $\sum v_r w_s$ satisfying the following properties for all $v \in V$, $w \in W$:

$$v(w_1 + w_2) = vw_1 + vw_2, \quad (v_1 + v_2)w = v_1 w + v_2 w,$$

$$\alpha(vw) = (\alpha v)w = v(\alpha w).$$

Furthermore, the tensor product is unique in the following sense: let Y be any complex vector space, and let B be any map $B : V \times W \to Y$ which is linear in both v and w. Then there exists a unique linear map $\widetilde{B} : V \otimes W \to Y$, such that $B = \widetilde{B} \circ i$.

We indicate how to build new representations from known ones.

3.4.10. Definitions. Let (π, V), (ρ, W) be representations of the group G.

(a) Let $V^* = \text{Hom}(V, \mathbb{C})$ be the vector space dual to V. The *conjugate* representation (π^*, V^*) of (π, V) is the representation of G on V^* defined by

$$(\pi^*(g) f)(x) = f(\pi(g^{-1})x) \qquad (g \in G, \; x \in V, \; f \in V^*).$$

(b) The *direct sum* of π and ρ is the representation $(\pi \oplus \rho, V \oplus W)$ of G on $V \oplus W$, defined by

$$(\pi \oplus \rho)(g)(v, w) = (\pi(g)v, \rho(g)w) \qquad (g \in G, \; v \in V, \; w \in W).$$

(c) The *tensor product* of π and ρ is the representation $(\pi \otimes \rho, V \otimes W)$ of G on $V \otimes W$, defined on elementary tensors $v \otimes w$ by

$$(\pi \otimes \rho)(g)(v \otimes w) = \pi(g)v \otimes \rho(g)w \qquad (g \in G, \; v \in V, \; w \in W).$$

3.4.11. Example. Let (π, V), (ρ, W) be representations of G. On $\text{Hom}(V, W)$, consider the representation σ defined by

$$\sigma(g)(T) = \rho(g) T \pi(g^{-1}) \qquad (g \in G, \; T \in \text{Hom}(V, W)).$$

Let us show that $\rho \otimes \pi^*$ is equivalent to σ. Indeed, for $f \in V^*$, $w \in W$, define a rank 1 operator $\theta_{w,f} \in \text{Hom}(V, W)$ by

$$\theta_{w,f}(v) = f(v) w \qquad (v \in V).$$

The map $B : W \times V^* \to \text{Hom}(V, W) : (w, f) \mapsto \theta_{w,f}$ is bilinear, so we get a linear map

$$\widetilde{B} : W \otimes V^* \to \text{Hom}(V, W) : w \otimes f \mapsto \theta_{w,f}.$$

The map $\widetilde{B}$ is onto, as one sees by taking bases for V and W, and the dual basis for V^*. Since $\dim_{\mathbb{C}}(W \otimes V^*) = \dim_{\mathbb{C}} \text{Hom}(V, W)$, the map $\widetilde{B}$ is an isomorphism. Finally, for $g \in G$, $w \in W$, $f \in V^*$, we have

$$\sigma(g)\theta_{w,f} = \theta_{\rho(g)w, \pi^*(g)f} = \widetilde{B}(\rho(g)w \otimes \pi^*(g)f),$$

so $\sigma(g)\widetilde{B} = \widetilde{B}(\rho \otimes \pi^*)(g)$, or $\widetilde{B} \in \text{Hom}_G(\rho \otimes \pi^*, \sigma)$.

Recall that a complex vector space has a hermitian scalar product $(\cdot \mid \cdot)$ satisfying the following properties:

(i) $(v \mid v) > 0$ if $v \neq 0$,

(ii) $(v \mid w) = \overline{(w \mid v)}$.

We now have the following results.

3.4.12. Proposition. Let (π, V) be a representation of the finite group G.

(i) There exists a hermitian scalar product $\langle \cdot \mid \cdot \rangle$ on V which is invariant under $\pi(G)$; that is, $\langle \pi(g) v_1 \mid \pi(g) v_2 \rangle = \langle v_1 \mid v_2 \rangle$ for every $g \in G$, $v_1, v_2 \in V$.

(ii) Every invariant subspace W of π admits an invariant complement; in other words, there exists an invariant subspace W', such that $W \cap W' = \{0\}$ and $W + W' = V$.

(iii) If $V \neq \{0\}$, then π is equivalent to a direct sum of irreducible representations of G.

Proof.

(i) Let $(\cdot \mid \cdot)$ be any hermitian scalar product on V. Set

$$\langle v_1 \mid v_2 \rangle = \sum_{h \in G} (\pi(h) v_1 \mid \pi(h) v_2) \qquad (v_1, v_2 \in V).$$

Then $\langle \cdot \mid \cdot \rangle$ is a hermitian scalar product, which is invariant, since, for $g \in G$,

$$\langle \pi(g) v_1 \mid \pi(g) v_2 \rangle = \sum_{h \in G} (\pi(hg) v_1 \mid \pi(hg) v_2)$$

$$= \sum_{h' \in G} (\pi(h') v_1 \mid \pi(h') v_2) = \langle v_1 \mid v_2 \rangle,$$

where the second equality follows from the change of variables $h' = hg$ in G.

(ii) Let W be an invariant subspace of π. By (i) under Proof, we may assume that π leaves invariant some hermitian scalar product $\langle \cdot \mid \cdot \rangle$. Define W' as the orthogonal of W with respect to $\langle \cdot \mid \cdot \rangle$:

$$W' = \{v \in V : \langle v \mid w \rangle = 0 \quad \forall w \in W\}.$$

W' is clearly a complement to W; let us check that W' is an invariant subspace. For $g \in G$, $v \in W'$, $w \in W$, we have

$$\langle \pi(g) v \mid w \rangle = \langle v \mid \pi(g^{-1}) w \rangle = 0,$$

since W is invariant and $\pi(g)$ is unitary. Thus, W' is also invariant under $\pi(G)$.

(iii) We prove the statement by induction on $\dim_{\mathbb{C}} V$. If $\dim_{\mathbb{C}} V = 1$, then π is irreducible (see Example 3.4.6 (i)). If $\dim_{\mathbb{C}} V > 1$, either π is irreducible, and there is nothing to prove, or π admits a nontrivial invariant subspace W. By (ii), we can find another nontrivial invariant subspace W' which is a complement to W. The map

$W \oplus W' \to V : (w, w') \mapsto w + w'$ realizes an equivalence between $\pi \mid_W \oplus \pi \mid_{W'}$ and π. Furthermore, by our induction assumption, the representations $\pi \mid_W$ and $\pi \mid_{W'}$ are equivalent to direct sums of irreducible representations. $\square$

For a representation (π, V) of G, we denote by

$$V^G = \{v \in V : \pi(g)v = v \quad \forall g \in G\}$$

the space of $\pi(G)$-fixed vectors in V. This is an invariant subspace of π.

3.4.13. Example. Let X be a finite G-space. A function $f \in \mathbb{C}X$ is fixed under $\lambda_X(G)$ if and only if f is constant on orbits of G in X. In particular, $\dim_{\mathbb{C}}(\mathbb{C}X)^G$ is the number of orbits of G in X.

3.4.14. Proposition. Let (π, V) be a representation of the finite group G. Set $P_\pi = \frac{1}{|G|} \sum_{g \in G} \pi(g)$. Then,

(i) $P_\pi^2 = P_\pi$; i.e., P_π is an idempotent in $\operatorname{End} V = \operatorname{Hom}(V, V)$;
(ii) for every $h \in G$, $\pi(h) P_\pi = P_\pi \pi(h) = P_\pi$;
(iii) $\operatorname{Im} P_\pi = V^G$;
(iv) $\frac{1}{|G|} \sum_{g \in G} \operatorname{Tr} \pi(g) = \dim_{\mathbb{C}}(V^G)$, where Tr denotes the trace of a matrix.

Proof.

(i) Noticing that, for fixed $s \in G$, there are $|G|$ pairs $(g, h) \in G \times G$, such that $gh = s$, we compute

$$P_\pi^2 = \frac{1}{|G|^2} \sum_{g \in G} \sum_{h \in G} \pi(gh) = \frac{1}{|G|} \sum_{s \in G} \pi(s) = P_\pi.$$

(ii)

$$\pi(h) P_\pi = \frac{1}{|G|} \sum_{g \in G} \pi(hg) = \frac{1}{|G|} \sum_{g' \in G} \pi(g') = P_\pi,$$

where the second equality is obtained by means of the change of variables $g' = hg$. The equality $P_\pi \pi(h) = P_\pi$ is proved in a similar way.

(iii) The image of an idempotent map is its fixed point set:

$$\operatorname{Im} P_\pi = \{v \in V : P_\pi(v) = v\}.$$

Clearly, from definition, if $v \in V^G$, then $P_\pi(v) = v$. Conversely, if $P_\pi(v) = v$, then by (ii), for every $h \in V$:

$$\pi(h)\, v = \pi(h)\, P_\pi(v) = P_\pi(v) = v.$$

(iv) The trace of an idempotent map is the dimension of its image. So, by (iii) and the linearity of the trace,

$$\frac{1}{|G|} \sum_{g \in G} \mathrm{Tr}\, \pi(g) = \mathrm{Tr}\, P_\pi = \dim_{\mathbb{C}}(V^G). \quad \square$$

In Proposition 3.4.14(iv), an important concept appears: the character of a representation.

3.4.15. Definition. Let (π, V) be a representation of G. The *character* of π is the function $\chi_\pi : G \to \mathbb{C} : g \mapsto \mathrm{Tr}\, \pi(g)$.

3.4.16. Example. Let X be a finite G-space. To compute χ_{λ_X}, we may use the basis of $\mathbb{C}X$ consisting of characteristic functions $(\delta_x)_{x \in X}$ of points. Since $\lambda_X(g)\, \delta_x = \delta_{gx}$ for $g \in G$, we see that $\lambda_X(g)$ is a permutation matrix. So the trace of $\lambda_X(g)$ is the number of 1's down the diagonal, or, equivalently, $\chi_{\lambda_X}(g)$ is the number of fixed points of g in X.

Specializing this to the case of the left regular representation of the finite group G, we get

$$\chi_{\lambda_X}(g) = \begin{cases} |G| & \text{if } g = 1 \\ 0 & \text{if } g \neq 1. \end{cases}$$

We study the behavior of the character under the constructions of representations in 3.4.10.

3.4.17. Proposition. Let (π, V), (ρ, W) be representations of G.

  (i) $\chi_{\pi^*}(g) = \chi_\pi(g^{-1})$ (for $g \in G$);
  (ii) $\chi_{\pi \oplus \rho} = \chi_\pi + \chi_\rho$;
  (iii) $\chi_{\pi \otimes \rho} = \chi_\pi\, \chi_\rho$;
  (iv) If π is equivalent to ρ, then $\chi_\pi = \chi_\rho$.

Proof. Let $e_1, \ldots, e_m$ (resp. $f_1, \ldots, f_n$) be a basis of V (resp. W).

  (i) Let $e_1^*, \ldots, e_m^*$ be the dual basis of $e_1, \ldots, e_m$. In this basis of V^*, the matrix of $\pi^*(g)$ is the transpose of the matrix of $\pi(g^{-1})$ in the basis $e_1, \ldots, e_m$. The result follows.

(ii)   In $V \oplus W$, we have $(\pi \oplus \rho)(g) = \begin{pmatrix} \pi(g) & O \\ O & \rho(g) \end{pmatrix}$.

(iii) Let $\pi(g)_{ik}$ (resp. $\rho(g)_{j\ell}$) be the matrix of $\pi(g)$ (resp. $\rho(g)$) in the given basis. Then the matrix of $\pi(g) \otimes \rho(g)$ in the basis $(e_i \otimes f_j)_{1 \le i \le m; 1 \le j \le n}$ of $V \otimes W$ is $(\pi(g)_{ik}\, \rho(g)_{j\ell})_{\substack{1 \le i, k \le m \\ 1 \le j, \ell \le n}}$. Then,

$$\chi_{\pi \otimes \rho}(g) = \sum_{i=1}^{m} \sum_{j=1}^{n} \pi(g)_{ii}\, \rho(g)_{jj} = \left(\sum_{i=1}^{m} \pi(g)_{ii} \right) \left(\sum_{j=1}^{n} \rho(g)_{jj} \right)$$

$$= \chi_\pi(g)\, \chi_\rho(g).$$

(iv)   If $T \in \mathrm{Hom}_G(\pi, \rho)$ is invertible, π and ρ are equivalent; then, by the trace property,

$$\chi_\rho(g) = \mathrm{Tr}(T\, \pi(g)\, T^{-1}) = \mathrm{Tr}\, \pi(g) = \chi_\pi(g). \quad \square$$

As a function on the group G, the character has the following properties.

3.4.18. Lemma. Let (π, V) be a representation of the finite group G.

(a) $\chi_\pi(1) = \dim_{\mathbb{C}} V$;
(b) $\chi_\pi(g) = \overline{\chi_\pi(g^{-1})}$ for $g \in G$;
(c) $\chi_\pi(g) = \chi_\pi(hg\, h^{-1})$ for $g, h \in G$.

Proof.
(a) This is clear, by definition.
(b) From Proposition 3.4.12(i), we know that $\pi(G)$ leaves invariant a hermitian scalar product $\langle \cdot \mid \cdot \rangle$ on V. Let $e_1, \ldots, e_m$ be an orthonormal basis of V, with respect to $\langle \cdot \mid \cdot \rangle$. Then

$$\chi_\pi(g^{-1}) = \sum_{i=1}^{m} \langle \pi(g^{-1}) e_i \mid e_i \rangle = \sum_{i=1}^{m} \langle e_i \mid \pi(g) e_i \rangle$$

$$= \sum_{i=1}^{m} \overline{\langle \pi(g) e_i \mid e_i \rangle} = \overline{\chi_\pi(g)}.$$

(c) We leave this as an exercise. $\square$

We now define the scalar product of two functions $f_1, f_2 : G \to \mathbb{C}$ as

$$\langle f_1 \mid f_2 \rangle_G = \frac{1}{|G|} \sum_{g \in G} f_1(g)\, \overline{f_2(g)}.$$

Referring to Schur's lemma (Theorem 3.4.9), we will show that the characters associated to irreducible representations of a group G form an orthonormal system with respect to this inner product.

3.4.19. Theorem. Let (π, V), (ρ, W) be representations of the finite group G. Then $\langle \chi_\rho \mid \chi_\pi \rangle_G = \dim_{\mathbb{C}} \operatorname{Hom}_G(\pi, \rho)$.

Proof. We compute

$$
\begin{aligned}
\langle \chi_\rho \mid \chi_\pi \rangle_G &= \frac{1}{|G|} \sum_{g \in G} \chi_\rho(g) \cdot \overline{\chi_\pi(g)} \\
&= \frac{1}{|G|} \sum_{g \in G} \chi_\rho(g)\, \chi_\pi(g^{-1}) && \text{(by Lemma 3.4.18(b))} \\
&= \frac{1}{|G|} \sum_{g \in G} \chi_\rho(g)\, \chi_{\pi^*}(g) && \text{(by Proposition 3.4.17(i))} \\
&= \frac{1}{|G|} \sum_{g \in G} \chi_{\rho \otimes \pi^*}(g) && \text{(by Proposition 3.4.17(iii))} \\
&= \frac{1}{|G|} \sum_{g \in G} \operatorname{Tr}(\rho \otimes \pi^*)(g) \\
&= \dim_{\mathbb{C}} (W \otimes V^*)^G && \text{(by Proposition 3.4.14(iv)).}
\end{aligned}
$$

By Example 3.4.11, the representation $\rho \otimes \pi^*$ is equivalent to the representation σ on $\operatorname{Hom}(V, W)$ defined by

$$
\sigma(g)(T) = \rho(g)\, T\, \pi(g^{-1}) \qquad (T \in \operatorname{Hom}(V, W),\ g \in G).
$$

Clearly, $T \in \operatorname{Hom}(V, W)$ is $\sigma(G)$-fixed if and only if T intertwines π and ρ, that is, if $T \in \operatorname{Hom}_G(\pi, \rho)$. So we see that

$$
\dim_{\mathbb{C}}(W \otimes V^*)^G = \dim_{\mathbb{C}} \operatorname{Hom}(V, W)^G = \dim_{\mathbb{C}} \operatorname{Hom}_G(V, W). \qquad \square
$$

3.4.20. Corollary. Let (π, V) be a representation of the finite group G, with $V \neq \{0\}$. Write $\pi = \rho_1 \oplus \cdots \oplus \rho_k$, with $\rho_1, \ldots, \rho_k$ irreducible representations of G. (This is possible, by Proposition 3.4.12(iii)). Let (ρ, W) be an irreducible representation of G. The number of those ρ_i's, which are equivalent to ρ, is equal to $\langle \chi_\pi \mid \chi_\rho \rangle_G$; in particular, this number does not depend on the chosen decomposition of π as a direct sum of irreducible representations.

Proof.

$$\langle \chi_\pi \mid \chi_\rho \rangle_G = \sum_{i=1}^{k} \langle \chi_{\rho_i} \mid \chi_\rho \rangle_G \qquad \text{(by Proposition 3.4.17(ii))}$$

$$= \sum_{i=1}^{k} \dim_{\mathbb{C}} \operatorname{Hom}_G(\rho, \rho_i) \qquad \text{(by Theorem 3.4.19).}$$

By Schur's lemma 3.4.9:

$$\dim_{\mathbb{C}} \operatorname{Hom}_G(\rho, \rho_i) = \begin{cases} 1 & \text{if } \rho \text{ is equivalent to } \rho_i \\ 0 & \text{if not.} \end{cases}$$

So $\langle \chi_\pi \mid \chi_\rho \rangle_G$ is indeed the number of ρ_i's equivalent to ρ. $\square$

This result gives a useful criterion for irreducibility.

3.4.21. Corollary. Let (π, V) be a representation of the finite group G, with $V \neq \{0\}$. The representation π is irreducible if and only if $\langle \chi_\pi \mid \chi_\pi \rangle_G = 1$.

Proof. If π is irreducible, then, by Theorems 3.4.19 and Schur's lemma,

$$\langle \chi_\pi \mid \chi_\pi \rangle_G = \dim_{\mathbb{C}} \operatorname{Hom}_G(\pi, \pi) = 1.$$

If π is not irreducible, then by Proposition 3.4.12(ii) we can write V as the direct sum of two nonzero invariant subspaces W, W':

$$V = W \oplus W'.$$

Then the maps $V \to V : (w, w') \mapsto (w, 0)$ and $V \to V : (w, w') \mapsto (0, w')$ are linearly independent elements in $\operatorname{Hom}_G(\pi, \pi)$, so that, by Theorem 3.4.19,

$$\langle \chi_\pi \mid \chi_\pi \rangle_G = \dim_{\mathbb{C}} \operatorname{Hom}_G(\pi, \pi) \geq 2. \quad \square$$

We can now prove the uniqueness, up to order, of the decomposition of a representation into irreducible components. First, let $V = W_1 \oplus \cdots \oplus W_r$ be a decomposition of V into irreducible G-invariant subspaces. Now let $\pi : G \to \operatorname{Aut}(V)$, so $\pi = \pi^{W_1} \oplus \cdots \oplus \pi^{W_r} = \pi_1 \oplus \cdots \oplus \pi_r$. By the properties of the trace,

$$\chi_\pi = \chi_1 + \cdots + \chi_r,$$

where $\chi_i = \chi_{\pi_i}$.

By what we have just proved

$$\langle \chi_i, \chi_\pi \rangle = \langle \chi_i, \chi_1 \rangle + \cdots + \langle \chi_i, \chi_r \rangle.$$

Hence, $\langle \chi_i, \chi_\pi \rangle$ is precisely the number of isomorphic copies of π_i which appears in the decomposition of π. However, that number is clearly independent of any particular decomposition, since

$$\langle \chi_i, \chi_\pi \rangle = 1/|G| \sum_{g \in G} \chi_i(g) \chi_\pi(g^{-1}).$$

To this extent the character χ_π determines the representation of π, and, in fact, π must contain exactly $\langle \chi_i, \chi_\pi \rangle$ copies of each irreducible component π_i. Thus, we have shown the following.

3.4.22. Theorem. Let $\pi : G \to \mathrm{Aut}(V)$ be a representation of G, and let $\pi = \sum n_j \pi_j$ be a decomposition of π for which the π_j are distinct irreducible representations of G. Then this decomposition is unique up to possible re-ordering of its components.

3.4.23. Theorem. Two representations with the same character are isomorphic, and any given irreducible representation appears with the same multiplicity in each.

From Theorem 3.4.19, the characters of inequivalent irreducible representations are orthogonal in $\mathbb{C}G$ with respect to the scalar product $\langle \cdot \mid \cdot \rangle_G$: this immediately implies that G has at most $|G|$ irreducible representations, up to equivalence. The following is known as the *degree formula*.

3.4.24. Corollary. Let $(\rho_1, W_1), \ldots, (\rho_h, W_h)$ be the list of irreducible representations of the finite group G, up to equivalence. Let $n_i = \dim_{\mathbb{C}} W_i$ be the degree of ρ_i. Then $|G| = \sum_{i=1}^{h} n_i^2$.

Proof. We decompose the left regular representation λ_G into irreducible representations. By Corollary 3.4.20, the representation ρ_i appears $\langle \chi_{\lambda_G} \mid \chi_{\rho_i} \rangle_G$ times in λ_G. But $\chi_{\lambda_G} = |G| \delta_1$, by Example 3.4.16, so that

$$\langle \chi_{\lambda_G} \mid \chi_{\rho_i} \rangle_G = \frac{1}{|G|} \sum_{g \in G} |G| \delta_1(g) \cdot \overline{\mathrm{Tr}\, \rho_i(g)} = \overline{\mathrm{Tr}\, \rho_i(1)} = n_i.$$

This means that $\chi_{\lambda_G} = \sum_{i=1}^{h} n_i \chi_{\rho_i}$. Evaluating at the identity of G gives the required formula. $\square$

The degree formula is useful in determining whether a list of irreducible representations is complete.

We close this section with a construction of irreducible representations.

3.4.25. Definition. A G-space X is 2-*transitive* if, for any two ordered pairs (x_1, y_1), (x_2, y_2) in $X \times X$, with $x_i \neq y_i$ ($i = 1, 2$), there exists $g \in G$, such that $g\, x_1 = x_2$ and $g\, y_1 = y_2$.

It is easy to check that, if the action of G on X is 2-transitive, it is also transitive. Now, let X be a finite G-space. We denote by λ_X^0 the restriction of λ_X to the co-dimension 1 subspace

$$W_0 = \left\{ f \in \mathbb{C}X : \sum_{x \in X} f(x) = 0 \right\},$$

already considered in Example 3.4.4.

3.4.26. Proposition. Let G be a finite group, and let X be a finite G-space which is 2-transitive. Then λ_X^0 is an irreducible representation of G.

Proof. We consider the diagonal action of G on $X \times X$, given by

$$g\,(x, y) = (gx, gy) \qquad (g \in G\,;\, x, y \in X).$$

Since X is 2-transitive, G has exactly two orbits on $X \times X$: the diagonal

$$\Delta = \{(x, x) : x \in X\}$$

and its complement $X \times X - \Delta$. So

$$
\begin{aligned}
2 = \dim_{\mathbb{C}} \left(\mathbb{C}\,(X \times X) \right)^G && \text{(by Example 3.4.13)} \\
= \frac{1}{|G|} \sum_{g \in G} \operatorname{Tr} \lambda_{X \times X}(g) && \text{(by Proposition 3.4.14(iv))} \\
= \frac{1}{|G|} \sum_{g \in G} \chi_{\lambda_{X \times X}}(g).
\end{aligned}
$$

Now consider the map

$$\varphi : \mathbb{C}X \otimes \mathbb{C}X \to \mathbb{C}\,(X \times X)$$

defined by

$$\varphi(f_1 \otimes f_2) = g_{f_1, f_2},$$

where

$$g_{f_1, f_2}(x, y) = f_1(x)\, f_2(y).$$

The map φ defines a canonical isomorphism from $\mathbb{C}X \otimes \mathbb{C}X$ onto $\mathbb{C}\,(X \times X)$, which intertwines $\lambda_X \otimes \lambda_X$ and $\lambda_{X \times X}$. Then $\lambda_X \otimes \lambda_X$ and $\lambda_{X \times X}$ are equivalent; hence, $\chi_{\lambda_{X \times X}} = \chi_{\lambda_X}^2$ by Proposition 3.4.17. This means that

$$2 = \frac{1}{|G|} \sum_{g \in G} \chi_{\lambda_X}(g)^2$$
$$= \langle \chi_{\lambda_X} \mid \chi_{\lambda_X} \rangle_G,$$

since χ_{λ_X} is real-valued, by Example 3.4.16. Now, as explained in Example 3.4.4, the representation λ_X decomposes as the direct sum of λ_X^0, and the one-dimensional trivial representation on the space of constant functions on X. Then, by Proposition 3.4.17(ii),

$$\chi_{\lambda_X} = 1 + \chi_{\lambda_X}^0,$$

so that

$$2 = \langle 1 + \chi_{\lambda_X}^0 \mid 1 + \chi_{\lambda_X}^0 \rangle_G$$
$$= \langle 1 \mid 1 \rangle_G + 2 \langle 1 \mid \chi_{\lambda_X}^0 \rangle_G + \langle \chi_{\lambda_X}^0 \mid \chi_{\lambda_X}^0 \rangle_G.$$

Since $\langle 1 \mid 1 \rangle_G = 1$, we get

$$1 = 2 \langle 1 \mid \chi_{\lambda_X}^0 \rangle_G + \langle \chi_{\lambda_X}^0 \mid \chi_{\lambda_X}^0 \rangle_G.$$

By Theorem 3.4.19, the numbers $\langle 1 \mid \chi_{\lambda_X}^0 \rangle_G$ and $\langle \chi_{\lambda_X}^0 \mid \chi_{\lambda_X}^0 \rangle_G$ are nonnegative integers, and $\langle \chi_{\lambda_X}^0 \mid \chi_{\lambda_X}^0 \rangle_G > 0$. This forces

$$\langle \chi_{\lambda_X}^0 \mid \chi_{\lambda_X}^0 \rangle_G = 1,$$

which, by Corollary 3.4.21, implies that λ_X^0 is irreducible. $\quad\square$

Exercises on Section 3.4

1. Let G be an abelian group. Show, using Schur's lemma, that every irreducible representation of G with finite degree, has degree 1.

2. Let π be the representation of $G = \mathbb{Z}$ on $\mathbb{C}^2$ given by $\pi(n) = \begin{pmatrix} 1 & n \\ 0 & 1 \end{pmatrix}$ ($n \in \mathbb{Z}$). Show that none of the three statements of Proposition 3.4.12 holds for π.

3. Let G be a finite abelian group, of order n.
   (a) Show that G has exactly n irreducible representations (up to equivalence), given by n homomorphisms $\chi_1, \dots, \chi_n$ from G to $\mathbb{C}^\times$. [Hint: combine exercise 1 with the degree formula 3.4.22.]
   (b) Using Theorem 3.4.19, show that $\chi_1, \dots, \chi_n$ are an orthonormal basis of $\mathbb{C}G$ for the scalar product $\langle \cdot \mid \cdot \rangle_G$.

(c) For $G = \mathbb{Z}/n\mathbb{Z}$, set $\omega = e^{\frac{2\pi i}{n}}$; for $a \in \mathbb{Z}/n\mathbb{Z}$, set

$$e_a(z) = \omega^{az} \qquad (z \in \mathbb{Z}/n\mathbb{Z}).$$

Show that the e_a's ($a \in \mathbb{Z}/n\mathbb{Z}$) are the n homomorphisms $G \to \mathbb{C}^\times$.

4. Let G be a finite group, and let π, ρ be representations of G with finite degree. Show that π is equivalent to ρ if and only if $\chi_\pi = \chi_\rho$. (In other words the character determines the representation.)

5. Let G be a finite group of order N, and let π be a representation of G of degree n. Show that, for every $g \in G$, the complex number $\chi_\pi(g)$ is algebraic, such that $|\chi_\pi(g)| \leq n$. [Hint: $\pi(g)$ is diagonalizable in a suitable basis, and its eigenvalues are N-th roots of 1 in $\mathbb{C}$.]

6. Let G be a finite group. A function $f \in \mathbb{C}G$ is a *class function* if it is constant on conjugacy classes of G; i.e.,

$$f(g h g^{-1}) = f(h) \qquad \forall g, h \in G.$$

Denote by $\mathrm{Cl}(G)$ the space of class functions on G, endowed with the scalar product $\langle \cdot \mid \cdot \rangle_G$.

(a) Show that the dimension of $\mathrm{Cl}(G)$ is the number of conjugacy classes of G.

(b) Let (ρ, V) be a representation of G. For $f \in \mathrm{Cl}(G)$, set $\rho(f) = \sum_{g \in G} f(g) \rho(g)$. Using Schur's lemma, show that if ρ is irreducible, then $\rho(f)$ is a scalar operator on V; actually

$$\rho(f) = \frac{1}{\dim_\mathbb{C} V} \sum_{g \in G} f(g) \chi_\rho(g) = \frac{|G|}{\dim_\mathbb{C} V} \langle \chi_\rho \mid \overline{f} \rangle_G.$$

(c) Let $\chi_1, \ldots, \chi_h$ be the characters of the irreducible representations of G. Show that the χ_i's are an orthonormal family in $\mathrm{Cl}(G)$ (i.e., $\langle \chi_i \mid \chi_j \rangle_G = \delta_{ij}$).

(d) Show that the χ_i's are a basis of $\mathrm{Cl}(G)$. [Hint: for $f \in \mathrm{Cl}(G)$ such that $\langle \chi_i \mid f \rangle_G = 0$ for $i = 1, \ldots, h$, show that $\lambda_G(\overline{f}) = 0$. Applying $\lambda_G(\overline{f})$ to the function $\delta_1 \in \mathbb{C}G$, show that $\overline{f} = 0$.]

(e) Show that the number of irreducible representations of G is equal to the number of conjugacy classes in G.

7. Let G be a finite group, and let X be a finite G-space. Show that the following properties are equivalent:

(i) χ is 2-transitive;

(ii) G has exactly two orbits $X \times X$;

(iii) λ_X^0 is an irreducible representation of G.

8. A *space with lines* is a set X endowed with a family $\mathcal{L}$ of subsets, called *lines*, satisfying the following properties:

(a) Every two distinct points of X belong to a unique line.

(b) Every line has at least two points.

(c) Every point belongs to at least two lines.

For example, take for X the n-dimensional affine (resp. projective) space over some field. Then X, with the set of all affine (resp. projective) lines, is a space with lines. Suppose that some group G acts by automorphisms on a space with lines X (i.e., X is a G-space and G leaves $\mathcal{L}$ invariant). Assume that X is finite.

(i) The *X-ray transform* $T : \ell^2(X) \to \ell^2(\mathcal{L})$ is defined by

$$(T f)(L) = \sum_{x \in L} f(x) \qquad (f \in \ell^2(X), \ L \in \mathcal{L})$$

(i.e., we "integrate" f on lines in X). Show that T intertwines the permutation representations λ_X and $\lambda_{\mathcal{L}}$; i.e., $T \in \text{Hom}_G(\lambda_X, \lambda_{\mathcal{L}})$.

(ii) Compute $T^* T : \ell^2(X) \to \ell^2(X)$ and deduce that T is injective. (This means that a function $f \in \ell^2(X)$ can be reconstructed from its integrals over all lines.)

(iii) Show that the number of orbits of G on X is at most the number of orbits of G on $\mathcal{L}$. [Hint: use injectivity of T.] In particular, if G acts transitively on $\mathcal{L}$, then it acts transitively on X.

(iv) (The aim of this exercise is to show that (iii) may fail for infinite spaces with lines; you should not try it if you do not know about hyperbolic geometry.) Let X be the union of the real hyperbolic plane and of its circle at infinity. A line in X is the union of a hyperbolic line and of its two points at infinity. Let G be the group of isometries of real hyperbolic plane. Check that X is a space with lines and that G has two orbits on X, but acts transitively on lines of X. This example is due to G. Valette.

3.5. Degrees of Representations of $\text{PSL}_2(q)$

The aim of this section is to prove the following result, going back to Frobenius [27].

3.5.1. Theorem. Let $q \geq 5$ be a prime. The degree of any nontrivial representation of $\text{PSL}_2(q)$ is at least $\frac{q-1}{2}$.

Let B be the "$ax + b$" group of $\mathbb{F}_q$, that is, the group of affine transformations

$$z \mapsto az + b \qquad (a \in \mathbb{F}_q^\times, \ b \in \mathbb{F}_q)$$

of $\mathbb{F}_q$. Viewing $\mathbb{F}_q$ as a B-space, we may form the permutation representation $\lambda_{\mathbb{F}_q}$ of B, as in Example 3.4.2(iii), and the subrepresentation $\lambda^0_{\mathbb{F}_q}$ on

$$W_0 = \left\{ f \in \mathbb{C}\,\mathbb{F}_q : \sum_{z \in \mathbb{F}_q} f(z) = 0 \right\}.$$

3.5.2. Lemma. The representation $\lambda^0_{\mathbb{F}_q}$ is an irreducible representation of B,
of degree $q - 1$.

Proof. By Proposition 3.4.24, it is enough to check that $\mathbb{F}_q$ is 2-transitive, as
a B-space. So let (x_1, x_2), (y_1, y_2) be two pairs in $\mathbb{F}_q \times \mathbb{F}_q$, with $x_1 \neq x_2$ and
$y_1 \neq y_2$. We need to find an affine map g, such that $g(x_i) = y_i$ $(i = 1, 2)$.
Geometrically, if we think of the graph of g as a subset of $\mathbb{F}_q \times \mathbb{F}_q$, we need
to find the line through the points (x_1, y_1) and (x_2, y_2), as in Figure 3.1.
But of course this line is given by

$$g(z) = \frac{y_2 - y_1}{x_2 - x_1}(z - x_1) + y_1$$

$(z \in \mathbb{F}_q)$, so that not only g exists, but it is unique. $\quad\square$

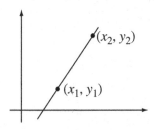

Figure 3.1

Recall that we denote by

$$\varphi : \text{SL}_2(q) \to \text{PSL}_2(q)$$

the canonical map and, as in the proof of Proposition 3.3.5, we let

$$B_0 = \varphi \left\{ \begin{pmatrix} a & b \\ 0 & a^{-1} \end{pmatrix} : a \in \mathbb{F}_q^\times, b \in \mathbb{F}_q \right\}$$

the stabilizer in PSL$_2(q)$ of $\infty \in P^1(\mathbb{F}_q)$. The action of B_0 on $\mathbb{F}_q$ is then given
by

$$\varphi_A(z) = a^2 z + ab, \qquad \text{where } A = \begin{pmatrix} a & b \\ 0 & a^{-1} \end{pmatrix}.$$

This means that B_0 identifies with a subgroup of index 2 in B. Actually, denoting by α the surjective homomorphism

$$\alpha : B \to \mathbb{F}_q^\times : (z \mapsto az + b) \mapsto a,$$

we see that $B_0 = \alpha^{-1}(\mathbb{F}_q^{\times 2})$, where $\mathbb{F}_q^{\times 2}$ denotes the group of squares in $\mathbb{F}_q^\times$. Now we describe the representation theory of B_0.

3.5.3. Proposition. Let q be an odd prime. Up to equivalence, there are $\frac{q+3}{2}$ irreducible representations of B_0, comprising

- $\frac{q-1}{2}$ group homomorphisms $B_0 \to \mathbb{C}^\times$, factoring through $\alpha \mid_{B_0}$;
- two inequivalent representations ρ_+, ρ_- of degree $\frac{q-1}{2}$.

Proof. Since $\mathbb{F}_q^{\times 2}$ is an abelian group of order $\frac{q-1}{2}$, by exercise 3 in section 3.4 it has exactly $\frac{q-1}{2}$ irreducible representations, all of degree 1, given by homomorphisms $\chi_1, \ldots, \chi_{\frac{q-1}{2}}$ from $\mathbb{F}_q^\times$ to $\mathbb{C}^\times$. Composing with $\alpha \mid_{B_0}$ gives $\frac{q-1}{2}$ homomorphisms $\chi_1 \circ \alpha \mid_{B_0}, \ldots, \chi_{\frac{q-1}{2}} \circ \alpha \mid_{B_0}$ from B_0 to $\mathbb{C}^\times$.

On the other hand, consider the restriction of $\lambda_{\mathbb{F}_q}^0$ to B_0: we are going to show that this restriction decomposes as the direct sum of two inequivalent, irreducible representations ρ_+, ρ_- of B_0, both of degree $\frac{q-1}{2}$. To see this, we appeal again to exercise 3 in section 3.4: set $\omega = e^{\frac{2\pi i}{q}}$ and consider, for $c \in \mathbb{F}_q$, the homomorphism,

$$e_c : \mathbb{F}_q \to \mathbb{C}^\times,$$

given by

$$e_c(z) = \omega^{cz}.$$

Then the e_c's, with $c \in \mathbb{F}_q$, are a basis of $\mathbb{C}\mathbb{F}_q$. In particular the e_c's, with $c \in \mathbb{F}_q^\times$, are a basis of the subspace W_0 from Proposition 3.4.26. Note that, if $g \in B$ is given by $g(z) = az + b$, then

$$(\lambda_{\mathbb{F}_q}^0(g)e_c)(z) = e_c(g^{-1}z) = e_c\left(\frac{z-b}{a}\right) = \omega^{-\frac{cb}{a}} \cdot \omega^{\frac{cz}{a}} = \omega^{-\frac{cb}{a}} e_{c/a}(z),$$

or else

$$\lambda_{\mathbb{F}_q}^0(g)e_c = \omega^{-\frac{cb}{a}} e_{c/a}.$$

Denote then by W_+ (resp. W_-) the subspace of W_0 generated by the e_c's, with c taking all square values (resp., nonsquare values) in $\mathbb{F}_q^\times$:

$$W_+ = \text{span} \langle e_c : c \in \mathbb{F}_q^{\times 2} \rangle$$
$$W_- = \text{span} \langle e_c : c \in \mathbb{F}_q^\times - \mathbb{F}_q^{\times 2} \rangle.$$

The preceding formula shows that W_+ and W_- are invariant subspaces for the restriction of $\lambda^0_{\mathbb{F}_q}$ to B_0; so we denote by ρ_+ (resp. ρ_-) the restriction of $\lambda^0_{\mathbb{F}_q} \mid_{B_0}$ to W_+ (resp. W_-). Note that

$$\dim_{\mathbb{C}} W_+ = \dim_{\mathbb{C}} W_- = \frac{q-1}{2},$$

so that ρ_+, ρ_- have degree $\frac{q-1}{2}$. To show that ρ_+, ρ_- are irreducible and inequivalent, we first observe that, if $g \in B - B_0$, then $\lambda^0_{\mathbb{F}_q}(g)$ exchanges W_+ and W_-. (This follows from the formula for the action of $\lambda^0_{\mathbb{F}_q}(g)$ on the e_c's.) This means that, in the decomposition $W_0 = W_+ \oplus W_-$, we have

$$\lambda^0_{\mathbb{F}_q}(g) = \begin{cases} \begin{pmatrix} \rho_+(g) & 0 \\ 0 & \rho_-(g) \end{pmatrix} & \text{if } g \in B_0\,; \\[2em] \begin{pmatrix} 0 & * \\ * & 0 \end{pmatrix} & \text{if } g \in B - B_0. \end{cases}$$

So, at the level of characters,

$$\chi_{\lambda^0_{\mathbb{F}_q}}(q) = \begin{cases} \chi_{\rho_+}(g) + \chi_{\rho_-}(g) & \text{if } g \in B_0\,; \\ 0 & \text{if } g \in B - B_0. \end{cases}$$

By Lemma 3.5.2, the representation $\lambda^0_{\mathbb{F}_q}$ of B, is irreducible. By Corollary 3.4.21, this means

$$
\begin{aligned}
1 &= \langle \chi_{\lambda^0_{\mathbb{F}_q}} \mid \chi_{\lambda^0_{\mathbb{F}_q}} \rangle_B = \frac{1}{|B|} \sum_{g \in B} |\chi_{\lambda^0_{\mathbb{F}_q}}(g)|^2 \\
&= \frac{1}{2|B_0|} \left[\sum_{g \in B_0} |\chi_{\lambda^0_{\mathbb{F}_q}}(g)|^2 + \sum_{g \in B - B_0} |\chi_{\lambda^0_{\mathbb{F}_q}}(g)|^2 \right] \\
&= \frac{1}{2|B_0|} \sum_{g \in B_0} |\chi_{\rho_+}(g) + \chi_{\rho_-}(g)|^2 \qquad \text{(by the previous formula)} \\
&= \frac{1}{2} \langle \chi_{\rho_+} + \chi_{\rho_-} \mid \chi_{\rho_+} + \chi_{\rho_-} \rangle_{B_0} \\
&= \frac{1}{2} \left[\langle \chi_{\rho_+} \mid \chi_{\rho_+} \rangle_{B_0} + 2 \operatorname{Re} \langle \chi_{\rho_+} \mid \chi_{\rho_-} \rangle_{B_0} + \langle \chi_{\rho_-} \mid \chi_{\rho_-} \rangle_{B_0} \right].
\end{aligned}
$$

By Theorem 3.4.19, the scalar products $\langle \chi_{\rho_+} \mid \chi_{\rho_+} \rangle_{B_0}$, $\langle \chi_{\rho_+} \mid \chi_{\rho_-} \rangle_{B_0}$ and $\langle \chi_{\rho_-} \mid \chi_{\rho_-} \rangle_{B_0}$ are nonnegative integers, with $\langle \chi_{\rho_+} \mid \chi_{\rho_+} \rangle_{B_0} > 0$ and $\langle \chi_{\rho_-} \mid \chi_{\rho_-} \rangle_{B_0} > 0$. This forces $\langle \chi_{\rho_+} \mid \chi_{\rho_+} \rangle_{B_0} = \langle \chi_{\rho_-} \mid \chi_{\rho_-} \rangle_{B_0} = 1$ and $\langle \chi_{\rho_+} \mid \chi_{\rho_-} \rangle_{B_0} = 0$. The first two equalities mean, by Corollary 3.4.21, that ρ_+ and ρ_- are irreducible representations of B_0; the latter means, by Theorem 3.4.9, that ρ_+ and ρ_- are not equivalent.

To show that this list of irreducible representations of B_0 is complete, we apply the degree formula 3.4.22: on the one hand, we have

$$|B_0| = \frac{q(q-1)}{2};$$

on the other hand, the sum of squares of degrees of irreducible representations obtained so far is

$$\frac{q-1}{2} \cdot 1^2 + 2 \cdot \left(\frac{q-1}{2}\right)^2 = \frac{q(q-1)}{2}.$$

Hence, the list is complete. $\square$

Proof of Theorem 3.5.1. Let π be a nontrivial representation of PSL$_2$(q) on $\mathbb{C}^n$. Consider the restriction $\pi|_{B_0}$. By Proposition 3.4.12, we may decompose it as a direct sum of irreducible representations of B_0, whose list is given in Proposition 3.5.3. Since $q \geq 5$, the group PSL$_2$(q) is simple, by Theorem 3.2.2, so $\pi|_{B_0}$ is a faithful representation of B_0, meaning that $\pi|_{B_0}(g) \neq I$ if $g \neq I$. Now it follows from Proposition 3.5.3 that representations of degree 1 of B_0 factor through the homomorphism $\alpha|_{B_0} \colon B_0 \to \mathbb{F}_q^{\times 2}$, so that they are all trivial on the commutator subgroup of B_0. This implies that at least one of the irreducible representations ρ_+, ρ_- must appear in $\pi|_{B_0}$, so that $n \geq \frac{q-1}{2}$. $\square$

Exercises on Section 3.5

1. Let G be a finite group, and let H be a subgroup of G. Let (π, V) be a finite-dimensional representation of H.
    (a) Set $W = \{f : G \to V : f(gh) = \pi(h^{-1})f(g) \quad \forall g \in G, h \in H\}$.
        Show that W is a complex vector space and that
        $$\dim_\mathbb{C} W = [G : H] \dim_\mathbb{C} V.$$
    (b) Define the *induced representation* $\mathrm{Ind}_H^G \pi$ of G on W by
        $$((\mathrm{Ind}_H^G \pi)(g)f)(x) = f(g^{-1}x) \qquad (g, x \in G; f \in W).$$
        Check that $\mathrm{Ind}_H^G \pi$ is a linear representation of G on W.

2. Show that the representation $\lambda_{\mathbb{F}_q}^0$ of B is induced from a non-trivial representation of degree 1, of the subgroup of translations $z \mapsto z + b$ ($b \in \mathbb{F}_q$).

3. Show that $N = \varphi\left\{\begin{pmatrix} 1 & b \\ 0 & 1 \end{pmatrix} \mid b \in \mathbb{F}_q\right\}$ is the commutator subgroup of B_0 for $q \geq 4$.

3.6. Notes on Chapter 3

3.2 It is a classical result of Jordan that, for a field K and $n \geq 2$, the group $PSL_n(K)$ is simple, with the exceptions of $PSL_2(\mathbb{F}_2)$ and $PSL_2(\mathbb{F}_3)$. This can be found in many books in group theory. As we already noticed in the proof of Theorem 3.2.2, the proof we give is special to dimension 2.

3.3 Our proof of Lemma 3.3.11 is patterned after Dickson's proof (see also [34] and [64]). Note that the same counting argument is used in the classification of the finite subgroups of $PSL_2(\mathbb{C})$.

3.4 An excellent introduction to the representation theory of finite groups, is given in Chapter 1 of Serre's book [60]. We tried not to duplicate it, by following a somewhat different route to the main results, appealing more to tensor products of representations.

3.5 The lower bound in Theorem 3.5.1 on degrees of nontrivial representations of $PSL_2(q)$ is actually sharp; this follows from the classification of irreducible representations of $PSL_2(q)$ (see [27] and [49]).

Chapter 4
The Graphs $X^{p,q}$

4.1. Cayley Graphs

Let G be a group (finite or infinite), and let S be a nonempty, finite subset of G. We assume that S is symmetric; that is, $S = S^{-1}$.

4.1.1. Definition. The *Cayley graph* $\mathcal{G}(G, S)$ is the graph with vertex set $V = G$ and edge set

$$E = \{\{x, y\} : x, y \in G \,;\, \exists s \in S : y = xs\}.$$

Hence, two vertices are adjacent if one is obtained from the other by right multiplication by some element of S. Note that, since S is symmetric, this adjacency relation is also symmetric, so the resulting graphs are undirected.

Examples.

(a) $G = \mathbb{Z}/6\mathbb{Z}$, $S = \{1, -1\}$.

(b) $G = \mathbb{Z}/6\mathbb{Z}$, $S = \{2, -2\}$.

(c) $G = \mathbb{Z}/6\mathbb{Z}$, $S = \{3\}$.

(d) $G = \mathbb{Z}/6\mathbb{Z}$, $S = \{2, -2, 3\}$.

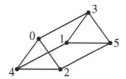

(e) $G = \text{Sym}(3)$, $S = \{(123), (132), (12)\}$.

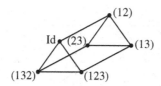

Examples (d) and (e) show that nonisomorphic groups can have isomorphic Cayley graphs.

(f) $G = \mathbb{Z}$, $S = \{1, -1\}$.

(g) $G = \mathbb{Z}$, $S = \{2, -2, 3, -3\}$.

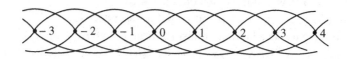

(h) $G = \mathbb{Z}^2$, $S = \{(1, 0), (-1, 0), (0, 1), (0, -1)\}$.

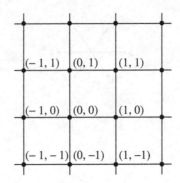

(i) $G = \mathbb{L}_2$, the free group on two generators a, b; $S = \{a, a^{-1}, b, b^{-1}\}$.

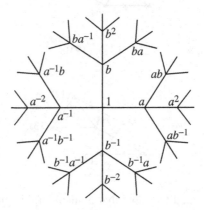

More generally, for the free group $\mathbb{L}_n$ on n symbols $a_1, \ldots, a_n$, with $S = \{a_1^{\pm 1}, \ldots, a_n^{\pm 1}\}$, the Cayley graph $\mathcal{G}(\mathbb{L}_n, S)$ is the $2n$-regular tree.

4.1.2. Proposition. Let $\mathcal{G}(G, S)$ be a Cayley graph; set $k = |S|$.

    (a) $\mathcal{G}(G, S)$ is a simple, k-regular, vertex-transitive graph.

    (b) $\mathcal{G}(G, S)$ has no loop if and only if $1 \notin S$.

    (c) $\mathcal{G}(G, S)$ is connected if and only if S generates G.

    (d) If there exists a homomorphism χ from G to the multiplicative group $\{1, -1\}$, such that $\chi(S) = \{-1\}$, then $\mathcal{G}(G, S)$ is bipartite. The converse holds provided $\mathcal{G}(G, S)$ is connected.

Proof.

(a) The adjacency matrix of $\mathcal{G}(G, S)$ is

$$A_{xy} = \begin{cases} 1 & \text{if there exists } s \in S \text{ such that } y = xs; \\ 0 & \text{otherwise.} \end{cases}$$

From this it is clear that $\mathcal{G}(G, S)$ is simple and k-regular. On the other hand, G acts on the left on $\mathcal{G}(G, S)$ by left multiplication: this action is transitive on $V = G$.

(b) This result is obvious.

(c) $\mathcal{G}(G, S)$ is connected if and only if every $x \in G$ is connected to $1 \in G$ by a path of edges. But this holds if and only if every $x \in G$ can be expressed as a word on the alphabet S, that is, if and only if S generates G.

(d) If the homomorphism $\chi : G \to \{\pm 1\}$ is given, then

$$V_{\pm} = \{x \in G : \chi(x) = \pm 1\}$$

defines a bipartition of $\mathcal{G}(G, S)$. For the converse, assume that $\mathcal{G}(G, S)$ is connected and bipartite. Denote by V_+ the class of the bipartition through $1 \in G$ and by V_- the other class. (Note that $S \subseteq V_-$.) We then define a map $\chi : G \to \{\pm 1\}$ by

$$\chi(x) = \begin{cases} 1 & \text{if } x \in V_+ \\ -1 & \text{if } x \in V_- \end{cases}.$$

To check that χ is a group homomorphism, we first observe that, since S generates G,

$$\chi(x) = (-1)^{\ell_S(x)},$$

where $\ell_S(x)$ is the word length of x with respect to S, hence, the distance from x to 1 in $\mathcal{G}(G, S)$. The fact that $G = V_+ \cup V_-$ then makes it clear that χ is a group homomorphism. $\square$

Exercises on Section 4.1

1. For the group G of quaternionic units (see exercise 3 in section 2.5) draw the Cayley graphs $\mathcal{G}(G, S)$ for the following choices of S:
   (a) $S = \{\pm i\}$;
   (b) $S = \{\pm i, \pm j\}$.

2. Let G be the group of symmetries of a square. Draw the Cayley graphs $\mathcal{G}(G, S)$ for the following choices of S:

(a) $S = \{s, r^{\pm 1}\}$, where s is a symmetry with respect to the median of a side, and r is a 90° rotation.

(b) $S = \{s, s'\}$, where s is as in (a), and s' is the symmetry with respect to a diagonal.

3. Let G be the group of rotations of a regular tetrahedron. Draw the Cayley graphs $\mathcal{G}(G, S)$ for the following choices of S:

(a) $S = \{s, r^{\pm 1}\}$, where s is the half-turn around the line joining the midpoints of two opposite edges and r is a rotation of angle 120° around the line joining a vertex to the center of the opposite face.

(b) $S = \{s, s'\}$ where s is as in (a), and s' is the half-turn around the line joining the midpoints of another pair of opposite edges.

(c) $S = \{r_1^{\pm 1}, r_2^{\pm 1}\}$, where r_i is a 120° rotation around the line joining a vertex to the center of the opposite face, and the axes of r_1, r_2 are distinct.

4. Let $\mathcal{G}(G, S)$ be a Cayley graph, with adjacency matrix A acting on $\ell^2(G)$. Denote by λ_G and ρ_G the left and right regular representations of G on $\ell^2(G)$ (see Example 3.4.2).

(a) Show that $A = \sum_{s \in S} \rho_G(s)$, as operators on $\ell^2(G)$.

(b) Let μ be an eigenvalue of A, with corresponding eigenspace V_μ. Show that V_μ is an invariant subspace of λ_G.

4.2. Construction of $X^{p,q}$

Let p, q be distinct odd primes. Recall from section 2.6 that we defined a distinguished set S_p of $p + 1$ integral quaternions of norm p.

We now consider reduction modulo q:

$$\tau_q : \mathbb{H}(\mathbb{Z}) \to \mathbb{H}(\mathbb{F}_q).$$

By Proposition 2.5.3 there exist integers x, y, such that $x^2 + y^2 + 1 \equiv 0$ (mod. q). Furthermore, by Proposition 2.5.2, any choice of such integers determines an isomorphism

$$\psi_q : \mathbb{H}(\mathbb{F}_q) \to M_2(\mathbb{F}_q)$$

enjoying the following two properties (see exercise 2 in section 2.5):

(a) $N(\alpha) = \det \psi_q(\alpha)$ for $\alpha \in \mathbb{H}(\mathbb{F}_q)$;
(b) if $\alpha \in \mathbb{H}(\mathbb{F}_q)$ is "real" (that is, if $\alpha = \bar{\alpha}$), then $\psi_q(\alpha)$ is a scalar matrix.

For $\alpha \in S_p$, we see that $\psi_q(\tau_q(\alpha))$ belongs to the invertible group $GL_2(q)$ of $M_2(\mathbb{F}_q)$, since $N(\alpha) = p \neq q$; also, $\psi_q(\tau_q(\alpha \bar{\alpha})) = \psi_q(\tau_q(\bar{\alpha} \alpha))$ is a nonzero scalar matrix in $GL_2(q)$. Now we compose further with the homomorphism

$$\varphi : GL_2(q) \to PGL_2(q)$$

(see section 3.1), whose kernel is exactly the subgroup of scalar matrices. We then set

$$S_{p,q} = (\varphi \circ \psi_q \circ \tau_q)(S_p).$$

The previous considerations show that $S_{p,q}$ is a symmetric subset of $PGL_2(q)$, so $S_{p,q}^{-1} = S_{p,q}$.

4.2.1. Lemma. If q is large enough with respect to p (for example, if $q > 2\sqrt{p}$), then $|S_{p,q}| = p + 1$.

Proof. Let $\alpha = a_0 + a_1 i + a_2 j + a_3 k$ and $\beta = b_0 + b_1 i + b_2 j + b_3 k$ be two distinct elements of S_p. Then, for some $i \in \{0, 1, 2, 3\}$, we have $a_i \neq b_i$. Since $N(\alpha) = N(\beta) = p$, we have $a_j, b_j \in (-\sqrt{p}, \sqrt{p})$ for every $j \in \{0, 1, 2, 3\}$; so if $q > 2\sqrt{p}$ we have $a_i \not\equiv b_i \pmod{q}$, and $\tau_q(\alpha) \neq \tau_q(\beta)$. Now set $A = (\psi_q \circ \tau_q)(\alpha)$ and $B = (\psi_q \circ \tau_q)(\beta)$, so that $A \neq B$ in $GL_2(q)$. Assume by contradiction that $\varphi_A = \varphi_B$ in $PGL_2(q)$. Then there exists $\lambda \in \mathbb{F}_q^\times$, such that $\lambda \neq 1$ and $A = \lambda B$. Taking determinants we get $p = \det A = \lambda^2 \det B = \lambda^2 p$; hence, $\lambda^2 = 1$, or $\lambda = -1$. From $A = -B$, we get $\alpha \equiv -\beta \pmod{q}$; i.e., $a_j \equiv -b_j \pmod{q}$ for every $j \in \{0, 1, 2, 3\}$. Since $q > 2\sqrt{p}$, we deduce $a_j = -b_j$, showing that $\alpha = -\beta$. By assumption, $a_0, b_0 \geq 0$, so $a_0 = b_0 = 0$; hence, $\beta = \overline{\alpha}$. But this contradicts the definition of S_p, since, as explained in section 2.6, if $\alpha \in S_p$ has $a_0 = 0$, then $\overline{\alpha} \notin S_p$. $\square$

The bound $2\sqrt{p}$ in Lemma 4.2.1 has nothing to do with the Ramanujan bound $2\sqrt{p}$ appearing in Theorem 4.2.2; this coincidence is just a numerical accident.

If p is a square modulo q, giving $\left(\frac{p}{q}\right) = 1$, then $S_{p,q}$ is actually contained in $PSL_2(q)$ (see exercise 3 in section 3.1). We define $X^{p,q}$ as the Cayley graph of $PSL_2(q)$ with respect to $S_{p,q}$:

$$X^{p,q} = \mathcal{G}(PSL_2(q), S_{p,q}).$$

If p is not a square modulo q, in which case $\left(\frac{p}{q}\right) = -1$, then $S_{p,q}$ is contained in $PGL_2(q) - PSL_2(q)$, and we define $X^{p,q}$ as the Cayley graph of $PGL_2(q)$ with respect to $S_{p,q}$:

$$X^{p,q} = \mathcal{G}(PGL_2(q), S_{p,q}).$$

The Holy Grail of this set of notes would be the following theorem.

4.2.2. Theorem. Let p, q be distinct, odd primes, with $q > 2\sqrt{p}$. The graphs $X^{p,q}$ are $(p+1)$-regular graphs which are connected and Ramanujan. Moreover,

(a) If $\left(\frac{p}{q}\right) = 1$, then $X^{p,q}$ is a nonbipartite graph with $\frac{q(q^2-1)}{2}$ vertices, satisfying the girth estimate

$$g(X^{p,q}) \geq 2 \log_p q.$$

(b) If $\left(\frac{p}{q}\right) = -1$, then $X^{p,q}$ is a bipartite graph with $q(q^2-1)$ vertices, satisfying $g(X^{p,q}) \geq 4 \log_p q - \log_p 4$.

4.2.3. Remark.

(a) The issue of connectedness of $X^{p,q}$ is a very important one that we address in section 4.3. By Proposition 4.1.2(c), it is equivalent to say that $S_{p,q}$ generates either $PSL_2(q)$ or $PGL_2(q)$, according to whether $\left(\frac{p}{q}\right) = 1$ or $\left(\frac{p}{q}\right) = -1$. This will be proved in section 4.3, under the slightly stronger assumption that $q > p^8$.

(b) Grails are seldom reached. We will not be able to prove the Ramanujan property for $X^{p,q}$ with our elementary means. We shall, however, indicate briefly how this property can be deduced from the Ramanujan conjecture on coefficients of modular forms. Nevertheless, we will prove by elementary means, in section 4.4, that for fixed p the family $(X^{p,q})_{q \text{ prime}}$ is a family of expanders, and we will get an explicit lower bound on the spectral gap.

(c) Some parts of Theorem 4.2.2 are easy to prove. It follows from Proposition 4.1.2(a) and Lemma 4.2.1 that $X^{p,q}$ is $(p+1)$-regular. The number of vertices of $X^{p,q}$ is given by Proposition 3.1.1, (b) and (c). If $\left(\frac{p}{q}\right) = -1$, the fact that $X^{p,q}$ is bipartite follows Proposition 4.1.2(d) and the group isomorphism $PGL_2(q)/PSL_2(q) \simeq \{\pm 1\}$.

Exercises on Section 4.2

1.   Show that $1 \notin S_{p,q}$, so that $X^{p,q}$ is a graph without loop.

2.   Construct a graph $Z^{p,q}$ as follows. The set of vertices is the projective line $P^1(\mathbb{F}_q)$, and the adjacency matrix is

$$A_{xy} = |\{s \in S_{p,q} : s(x) = y\}|$$

(for $x, y \in P^1(\mathbb{F}_q)$). Taking Theorem 4.2.2 for granted, show that $Z^{p,q}$ is a $(p+1)$-regular, connected, Ramanujan graph. [Hint: show that the spectrum of $Z^{p,q}$ is contained in the spectrum of $X^{p,q}$.] It may happen that $Z^{p,q}$ has loops or multiple edges: see the pictures of $Z^{5,13}$ and $Z^{5,17}$ in [57]. These pictures also show that $Z^{p,q}$ is not necessarily vertex-transitive. It might, however, be interesting to know that there are $(p+1)$-regular Ramanujan graphs on $q+1$ vertices.

4.3. Girth and Connectedness

In this section, we introduce another family $Y^{p,q}$ of $(p+1)$-regular graphs that will ultimately turn out to be isomorphic to $X^{p,q}$. Because the $Y^{p,q}$'s are defined as quotients of trees, it will be fairly easy to estimate their girth. We will see in section 4.4 that they are also tractable for spectral estimates.

Let p be an odd prime. Recall that after Theorem 2.5.13 we defined a subset Λ' of $\mathbb{H}(\mathbb{Z})$ as

$$\Lambda' = \{\alpha \in \mathbb{H}(\mathbb{Z}) : \alpha \equiv 1(\text{mod. } 2) \text{ or} \alpha \equiv i + j + k \ (\text{mod. } 2),$$
$$N(\alpha) \text{a power of} p\}.$$

On Λ', we define the following equivalence relation: $\alpha \sim \beta$ if there exist $m, n \in \mathbb{N}$, such that $p^m \alpha = \pm p^n \beta$. We denote by $[\alpha]$ the equivalence class of $\alpha \in \Lambda'$, by $\Lambda = \Lambda'/\sim$ the set of equivalence classes and by

$$Q : \Lambda' \to \Lambda$$

the quotient map $Q(\alpha) = [\alpha]$.

Note that $\sim$ is compatible with multiplication; that is, if $\alpha_1 \sim \beta_1, \alpha_2 \sim \beta_2$, then $\alpha_1 \alpha_2 \sim \beta_1 \beta_2$. Thus, Λ comes equipped with an associative product with unit.

Recall that before Definition 2.6.12 we defined a set

$$S_p = \{\alpha_1, \overline{\alpha_1}, \ldots, \alpha_s, \overline{\alpha_s}, \beta_1, \ldots, \beta_t\}$$

of $p+1$ integral quaternions of norm p, where α_i has $a_0^{(i)} > 0$ and β_j has $b_0^{(j)} = 0$. By definition $S_p \subset \Lambda'$.

4.3.1. Proposition.

(a) Λ is a group.

(b) The Cayley graph $\mathcal{G}(\Lambda, Q(S_p))$ is the $(p+1)$-regular tree.

Proof.

(a) For $\alpha \in \Lambda' : \alpha \overline{\alpha} = \overline{\alpha} \alpha \sim 1$; hence, $[\alpha]^{-1} = [\alpha]$, so Λ is a group.

(b) For $\alpha, \beta \in S_p$, one sees that $\alpha \sim \beta$ implies $\alpha = \beta$. So $|Q(S_p)| = p+1$.

By the existence part of Corollary 2.6.14, any $\alpha \in \Lambda'$ is equivalent to a reduced word over S_p; in other words, Λ is generated by $Q(S_p)$, and, by Proposition 4.1.2, the graph $\mathcal{G}(\Lambda, Q(S_p))$ is $(p+1)$-regular and connected. To prove that it is a tree, we have to show that it does not contain any circuit. So suppose by contradiction that it does contain a circuit $x_0, x_1, x_2, \ldots, x_{n-1}, x_g = x_0$ of length $g \geq 3$. By vertex-transitivity, we may assume $x_0 = [1]$. By definition of a Cayley graph, we have $x_1 = [\gamma_1]$, $x_2 = [\gamma_1 \gamma_2], \ldots, x_g = [\gamma_1 \gamma_2 \cdots \gamma_g]$ for some $\gamma_1, \gamma_2, \ldots, \gamma_g \in S_p$. Since $x_{k-1} \neq x_{k+1}$ for $1 \leq k \leq n-1$, the word $\gamma_1 \gamma_2 \ldots \gamma_g$ over S_p is reduced; i.e., it contains no occurrence of $\alpha_i \overline{\alpha_i}, \overline{\alpha_i} \alpha_i$ or β_j^2 ($1 \leq i \leq s; 1 \leq j \leq t$). The equality $[1] = [\gamma_1 \gamma_2 \ldots \gamma_g]$ in Λ becomes, in Λ',

$$p^m = \pm p^n \gamma_1 \gamma_2 \cdots \gamma_g.$$

But since $\gamma_1 \gamma_2 \ldots \gamma_g$ is a nontrivial reduced word over S_p, this contradicts the uniqueness part in Corollary 2.6.14, and the proof is complete. $\square$

As in section 4.2, we consider reduction modulo q:

$$\tau_q : \mathbb{H}(\mathbb{Z}) \rightarrow \mathbb{H}(\mathbb{F}_q);$$

which sends Λ' to the group $\mathbb{H}(\mathbb{F}_q)^\times$ of invertible elements in $\mathbb{H}(\mathbb{F}_q)$. Let Z_q be the following central subgroup of $\mathbb{H}(\mathbb{F}_q)^\times$:

$$Z_q = \{\alpha \in \mathbb{H}(\mathbb{F}_q)^\times : \alpha = \overline{\alpha}\}.$$

Let $\alpha, \beta \in \Lambda'$: if $\alpha \sim \beta$, then $\tau_q(\alpha)^{-1} \tau_q(\beta) \in Z_q$. This means that $\tau_q : \Lambda' \rightarrow \mathbb{H}(\mathbb{F}_q)^\times$ descends to a well-defined group homomorphism

$$\Pi_q : \Lambda \rightarrow \mathbb{H}(\mathbb{F}_q)^\times / Z_q.$$

We denote the kernel of Π_q by $\Lambda(q)$ and we identify the image of Π_q with the quotient group $\Lambda / \Lambda(q)$. We set $T_{p,q} = (\Pi_q \circ Q)(S_p)$.

One sees as in Lemma 4.2.1 that, for q sufficiently large with respect to p (for example, $q > 2\sqrt{p}$), one has $|T_{p,q}| = p + 1$. We define the graph $Y^{p,q}$ as the Cayley graph of $\Lambda / \Lambda(q)$ with respect to $T_{p,q}$:

$$Y^{p,q} = \mathcal{G}(\Lambda / \Lambda(q), T_{p,q}).$$

Since Λ is generated by $Q(S_p)$ (see Proposition 4.3.1), it follows from Proposition 4.1.2 that, for $q > 2\sqrt{p}$, the graph $Y^{p,q}$ is $(p+1)$-regular and connected.

Notice now that the isomorphism $\psi_q : \mathbb{H}(\mathbb{F}_q)^\times \rightarrow GL_2(q)$ of Proposition 2.5.2 sends Z_q to the subgroup of scalar matrices in $GL_2(q)$, which, in turn, form the kernel of $\varphi : \mathrm{GL}_2(q) \to \mathrm{PGL}_2(q)$. Hence, ψ_q descends to an isomorphism

$$\beta : \mathbb{H}(\mathbb{F}_q)^\times / Z_q \to \mathrm{PGL}_2(q).$$

This allows us to compare, by means of a commutative diagram, the constructions of $X^{p,q}$ and $Y^{p,q}$:

$$
\begin{array}{ccccc}
S_p \subset \Lambda' & \overset{\tau_q}{\longrightarrow} & \mathbb{H}(\mathbb{F}_q)^\times & \overset{\psi_q}{\longrightarrow} & \mathrm{GL}_2(q) \\
\downarrow{\scriptstyle \varrho} & & \downarrow & & \downarrow{\scriptstyle \varphi} \\
\Lambda & \underset{\Pi_q}{\longrightarrow} & \mathbb{H}(\mathbb{F}_q)^\times / Z_q & \underset{\beta}{\longrightarrow} & \mathrm{PGL}_2(q)
\end{array}
$$

(Here all vertical arrows are quotient maps.) The graph $X^{p,q}$ is defined by means of $\varphi \circ \psi_q \circ \tau_q$, while $Y^{p,q}$ is defined by means of $\Pi_q \circ Q$. We do not know yet that $X^{p,q}$ is connected, but we know exactly from which group it comes, namely, from either $\mathrm{PSL}_2(q)$ or $\mathrm{PGL}_2(q)$, depending whether or not p is a square modulo q. By contrast, $Y^{p,q}$ is connected by definition, but we did not identify the group $\Lambda/\Lambda(q)$ from which it comes. (For example, what is the order of $\Lambda/\Lambda(q)$?). However, since $\beta(T_{p,q}) = S_{p,q}$, we see that $Y^{p,q}$ is a connected component of $X^{p,q}$. Playing both constructions against each other, we will eventually see that $X^{p,q}$ is connected for $q > p^8$, so that $X^{p,q}$ is isomorphic to $Y^{p,q}$.

We first need to identify the "congruence subgroup" $\Lambda(q)$.

4.3.2. Lemma.

$$\Lambda(q) = \{[\alpha] \in \Lambda : \alpha = a_0 + a_1 i + a_2 j + a_3 k, q \mid a_1, a_2, a_3\}.$$

Proof.

$$
\begin{aligned}
[\alpha] \in \Lambda(q) &\Leftrightarrow \tau_q(\alpha) \in Z_q \\
&\Leftrightarrow q \text{ does not divide } a_0 \text{ and } q \mid a_1, a_2, a_3 \\
&\Leftrightarrow q \mid a_1, a_2, a_3,
\end{aligned}
$$

where the equivalence between the second and third lines follows from the fact that $N(\alpha)$ is a power of p, and $p \neq q$. $\square$

We can now give a lower bound for the girth of $Y^{p,q}$.

4.3.3. Proposition. One has $g(Y^{p,q}) \geq 2\log_p q$. If $\left(\frac{p}{q}\right) = -1$, we have the better inequality $g(Y^{p,q}) \geq 4\log_p q - \log_p 4$.

Proof. For simplicity's sake, write g for $g(Y^{p,q})$. Let $x_0, x_1, \ldots, x_{g-1}, x_g = x_0$ be the vertices of a circuit of length g in $Y^{p,q}$. By vertex-transitivity of $Y^{p,q}$ (see Proposition 4.1.2), we may assume that $x_0 = x_g = 1$ in $\Lambda/\Lambda(q)$. Since $Y^{p,q}$ is a Cayley graph, we find $t_1, \ldots, t_g \in T_{p,q}$, such that

$$x_i = t_1 t_2 \ldots t_i \qquad (1 \leq i \leq g).$$

Now $t_i = \Pi_q([\gamma_i])$ for a unique $\gamma_i \in S_p$ $(i = 1, \ldots, g)$. Write $\alpha = \gamma_1 \ldots \gamma_g \in \Lambda'$, with $\alpha = a_0 + a_1 i + a_2 j + a_3 k$. Then α is a reduced word over S_p, and $[\alpha] = [\gamma_1] \ldots [\gamma_g]$ is distinct from $[1]$ in Λ, by Proposition 4.3.1(b). So α is not equivalent to 1 in Λ', which implies that at least one of a_1, a_2, a_3 is nonzero. On the other hand,

$$\Pi_q([\alpha]) = t_1 t_2 \ldots t_g = x_g = 1,$$

so that $[\alpha] \in \Lambda(q)$. By Lemma 4.3.2, the prime q must divide a_1, a_2, a_3. Since one of them is nonzero, we get

$$p^g = N(\alpha) = a_0^2 + a_1^2 + a_2^2 + a_3^2 \geq q^2.$$

Taking logarithms in base p, we get the first statement.

Suppose now that $\left(\frac{p}{q}\right) = -1$. Since $p^g \equiv a_0^2 \pmod{q}$, we have

$$1 = \left(\frac{p^g}{q}\right) = \left(\frac{p}{q}\right)^g = (-1)^g,$$

so that g is even, say $g = 2h$. Now actually

$$p^{2h} \equiv a_0^2 \pmod{q^2}.$$

From exercise 1 in this section, it follows that

$$p^h \equiv \pm a_0 \pmod{q^2}.$$

On the other hand, $a_0^2 \leq p^g$, so $|a_0| \leq p^h$. Assume by contradiction that $g < 4\log_p q - \log_p 4 = \log_p \frac{q^4}{4}$, so $p^h < \frac{q^2}{2}$. Then $|p^h \mp a_0| < q^2$ and, from the previous congruence, we get $p^h = \pm a_0$. Then $p^g = a_0^2$, which forces $a_1 = a_2 = a_3 = 0$; this gives a contradiction. $\square$

4.3.4. Remark. From exercise (3) in section 1.3, we have, for $p \geq 5$,

$$g(Y^{p,q}) \leq 2 + 2\log_p |Y^{p,q}|;$$

therefore, from Proposition 4.3.3,

$$|Y^{p,q}| \geq \frac{q}{p}.$$

Better, if $\left(\frac{p}{q}\right) = -1$,

$$|Y^{p,q}| \geq \frac{q^2}{2p}.$$

This shows that $|Y^{p,q}| = |\Lambda/\Lambda(q)|$ grows at least linearly with q.

Here now is the main result of this section.

4.3.5. Theorem. Assume $p \geq 5$. For $q > p^8$, the graph $X^{p,q}$ is connected; therefore, $X^{p,q}$ is isomorphic to $Y^{p,q}$.

Proof. By Proposition 4.1.2(c), we have to show that $S_{p,q}$ generates $\mathrm{PSL}_2(q)$ if $\left(\frac{p}{q}\right) = 1$; and $\mathrm{PGL}_2(q)$, if $\left(\frac{p}{q}\right) = -1$. Recall the isomorphism $\beta : \mathbb{H}(\mathbb{F}_q)^\times / Z_q \to \mathrm{PGL}_2(q)$. Since $\beta(T_{p,q}) = S_{p,q}$, it is equivalent to prove

$$\beta\left(\Lambda/\Lambda(q)\right) = \begin{cases} \mathrm{PSL}_2(q) & \text{if } \left(\frac{p}{q}\right) = 1; \\ \mathrm{PGL}_2(q) & \text{if } \left(\frac{p}{q}\right) = -1. \end{cases}$$

In the second case, we already observed that $S_{p,q} \subset \mathrm{PGL}_2(q) - \mathrm{PSL}_2(q)$. Set $H_{p,q} = \mathrm{PSL}_2(q) \cap \beta\left(\Lambda/\Lambda(q)\right)$. We are left to prove that, in both cases,

$$H_{p,q} = \mathrm{PSL}_2(q).$$

In view of Theorem 3.3.4, this will follow from two facts: $|H_{p,q}| > 60$, and $H_{p,q}$ is not metabelian.

To prove that $|H_{p,q}| > 60$, we observe that, since $q > p^8$ and $p \geq 5$, we certainly have, from Remark 4.3.4,

$$|\Lambda/\Lambda(q)| \geq \frac{q}{p} > 120;$$

hence, $|H_{p,q}| > 60$.

To prove that $H_{p,q}$ is not metabelian, by exercise 2 in section 3.3, we must show that there exist g_1, g_2, g_3, g_4 in $H_{p,q}$, such that

$$[[g_1, g_2], [g_3, g_4]] \neq 1.$$

For that, we examine each case:

(a) If $\left(\frac{p}{q}\right) = 1$, we choose the g_i's as follows from among the elements of $S_{p,q}$: take for g_1 any element in $S_{p,q}$; choose g_2 to be distinct from $g_1^{\pm 1}$. Then take $g_3 = g_1$, and let $g_4 \notin \{g_1^{\pm 1}, g_2^{\pm 1}\}$. With this choice $[[g_1, g_2], [g_3, g_4]]$ is a reduced word of length 16 over $S_{p,q}$. By Proposition 4.3.3, the girth of $Y^{p,q}$ satisfies

$$g(Y^{p,q}) \geq 2\log_p q > 16;$$

as a consequence, any reduced word of length 16 over $S_{p,q}$ cannot be equal to 1 in $H_{p,q}$, since this would create a circuit of length at most 16 in $Y^{p,q}$.

(b) If $\left(\frac{p}{q}\right) = -1$, we first choose h_1, h_2, h_3 in $S_{p,q}$ as follows: let h_1 be any element of $S_{p,q}$; let h_2 be distinct from $h_1^{\pm 1}$ and $h_3 \notin \{h_1^{\pm 1}, h_2^{\pm 1}\}$. Then we set $g_1 = h_1 h_3$, $g_2 = h_2 h_3$, $g_3 = h_1 h_2$, and $g_4 = h_3 h_2$: these are elements of $H_{p,q}$. Then $[g_1, g_2] = h_1 h_3 h_2 h_1^{-1} h_3^{-1} h_2^{-1}$ and $[g_3, g_4] = h_1 h_2 h_3 h_1^{-1} h_2^{-1} h_3^{-1}$. Then $[[g_1, g_2], [g_3, g_4]]$ is a reduced word of length 24 on $S_{p,q}$. By Proposition 4.3.3, the girth of $Y^{p,q}$ satisfies

$$g(Y^{p,q}) \geq 4\log_p q - \log_p 4 > 24.$$

The conclusion then follows, as in (a). □

We summarize what this means for the graphs $X^{p,q}$ (in comparison to the statement of Theorem 4.2.2).

4.3.6. Corollary. Assume that $q > p^8$. The graphs $X^{p,q}$ are $(p+1)$-regular connected graphs. Moreover,

(a) If $\left(\frac{p}{q}\right) = 1$, then $X^{p,q}$ is nonbipartite, with

$$g(X^{p,q}) \geq \frac{2}{3} \log_p |X^{p,q}|.$$

(b) If $\left(\frac{p}{q}\right) = -1$, then $X^{p,q}$ is bipartite, with

$$g(X^{p,q}) \geq \frac{4}{3} \log_p |X^{p,q}| - \log_p 4.$$

Proof. Connectedness was established in Theorem 4.3.5. The girth estimates follow from Proposition 4.3.3 and the fact that $q^3 \geq |X^{p,q}|$. (See Proposition 3.1.1.) Assume that $\left(\frac{p}{q}\right) = 1$: in view of Proposition 4.2.1(d) and the connectedness of $X^{p,q}$, the fact that $X^{p,q}$ is not bipartite follows from the simplicity of $\text{PSL}_2(q)$ proved in Theorem 3.2.2. Finally, if $\left(\frac{p}{q}\right) = -1$, the fact that $X^{p,q}$ is bipartite was already observed in Remark 4.2.3(c). □

4.3.7. Remark. Recall from Definition 0.12 that a family $(X_m)_{m \geq 1}$ of finite, connected, k-regular graphs, with $|X_m| \to +\infty$ for $m \to +\infty$, has *large girth* if there exists $C > 0$, such that $g(X_m) \geq (C + o(1)) \log_{k-1} |X_m|$. From exercise (2) in section 1.3, we necessarily have $C \leq 2$. As already mentioned, Erdös and Sachs [25] gave a nonconstructive proof of such families with $C = 1$. For $\left(\frac{p}{q}\right) = -1$, the graphs $X^{p,q}$ are an explicit family of $(p+1)$-regular graphs with large girth, namely, $C = \frac{4}{3}$. This is one of the few examples in graph theory where explicit methods give better results than nonconstructive ones.

Exercises on Section 4.3

1. Let q be an odd prime; let a, b be integers, not divisible by q, such that $a^2 \equiv b^2$ (mod. q^2); show that $a \equiv \pm b$ (mod. q^2).

2. Show that, if $p \equiv 1$ (mod. 4) and $\left(\frac{p}{q}\right) = -1$, then $g(Y^{p,q}) \geq 4 \log_p q$.
    [Hint: in the proof of Proposition 4.3.3, observe that the congruence $p^h \equiv \pm a_0$ (mod. q^2) improves to $p^h \equiv \pm a_0$ (mod. $2q^2$), since p and a_0 are odd.]

3. Assume $p \equiv 1$ (mod. 4). Show that $\Lambda(2q) = \{[\alpha] \in \Lambda(2) : \alpha = a_0 + a_1 i + a_2 j + a_3 k, 2q \mid a_1, a_2, a_3\}$ (compare with Lemma 4.3.2).

4. How should one modify Theorem 4.3.5 (and its proof) to include the case $p = 3$?

5. (This exercise assumes some acquaintance with free groups.)
    (a) Assume that $p \equiv 1$ (mod. 4); show that the group $\Lambda(2)$ is isomorphic to the free group $\mathbb{L}_{\frac{p+1}{2}}$ on $\frac{p+1}{2}$ generators.
    (b) Let $\mathbb{L}_n$ be the free group on $a_1, \ldots, a_n$; set $S = \{a_1, \ldots, a_n, a_1^{-1}, \ldots, a_n^{-1}\}$. Let N be a normal subgroup of $\mathbb{L}_n$; denote by $\Pi : \mathbb{L}_n \to \mathbb{L}_n/N$ the quotient map, and assume that $\Pi |_S$ is one-to-one. Show that the girth of the Cayley graph $\mathcal{G}(\mathbb{L}_n/N, \Pi(S))$ is the minimum of the word length of the nonidentity elements in N. (When $p \equiv 1$ (mod. 4), this applies to the graphs $Y^{p,q}$ in Proposition 4.3.3.)

6. (This exercise requires some basic knowledge about free products.) Write $S_p = \{\alpha_1, \overline{\alpha_1}, \ldots, \alpha_s, \overline{\alpha_s}, \beta_1, \ldots, \beta_t\}$, as immediately before Proposition 4.3.1. Observe that in $\Lambda(2) : [\beta_j]^2 = [1]$ $(1 \leq j \leq t)$. Deduce that

$\Lambda(2)$ is isomorphic to the free product of $\mathbb{L}_s$ with t copies of $\mathbb{Z}/2\mathbb{Z}$:

$$\Lambda(2) \simeq \mathbb{L}_s * \underbrace{\mathbb{Z}/2\mathbb{Z} * \cdots * \mathbb{Z}/2\mathbb{Z}}_{t \text{ factors}}.$$

4.4. Spectral Estimates

In this section we prove that, for fixed p, the family $X^{p,q}$ is a family of expanders, with an explicit lower bound on the spectral gap when q is large enough with respect to p.

We shall denote by n the number of vertices of $X^{p,q}$, computed in Proposition 3.1.1, and by

$$\mu_0 = p + 1 > \mu_1 \geq \mu_2 \geq \cdots \geq \mu_{n-1}$$

the spectrum of its adjacency matrix. Recall from section 1.4 that f_m is the number of paths of length m without backtracking, starting, and ending at 1 on $X^{p,q}$. Because, by Proposition 4.1.2(a), the graph $X^{p,q}$ is vertex-transitive, the trace formula in Corollary 1.4.7 takes the following form for $X^{p,q}$:

$$\sum_{0 \leq r \leq \frac{m}{2}} f_{m-2r} = \frac{p^{\frac{m}{2}}}{n} \sum_{j=0}^{n-1} U_m \left(\frac{\mu_j}{2\sqrt{p}} \right),$$

for every $m \in \mathbb{N}$.

Our first task is to reinterpret the left-hand side of this trace formula. For this, we introduce the quadratic form in four variables:

$$Q(x_0, x_1, x_2, x_3) = x_0^2 + q^2(x_1^2 + x_2^2 + x_3^2),$$

and, for $m \geq 1$, we set

$$s_Q(p^m) = |\{(x_0, x_1, x_2, x_3) \in \mathbb{Z}^4 : Q(x_0, x_1, x_2, x_3) = p^m, \text{ either } x_0$$
$$\text{odd and } x_1, x_2, x_3 \text{ even, or } x_0 \text{ even and } x_1, x_2, x_3 \text{ odd}\}|.$$

4.4.1. Remark. Suppose either m even or $p \equiv 1$ (mod. 4). By reducing modulo 4, one sees in the previous definition that all the 4-tuples (x_0, x_1, x_2, x_3) appearing have x_0 odd and x_1, x_2, x_3 even. We introduce the quadratic form:

$$Q'(x_0, x_1, x_2, x_3) = x_0^2 + 4q^2(x_1^2 + x_2^2 + x_3^2);$$

then $s_Q(p^m)$ is exactly the number of integral representations of p^m by the quadratic form Q'.

We now come back to a general p.

4.4.2. Lemma. For $m \in \mathbb{N} : s_Q(p^m) = 2 \sum_{0 \le r \le \frac{m}{2}} f_{m-2r}$.

Proof. We identify $X^{p,q}$ with $Y^{p,q}$, by Theorem 4.3.5. Let $x_0 = 1, x_1, \ldots, x_{\ell-1}$, $x_\ell = 1$ be the vertices of a path of length ℓ, without backtracking, starting, and ending at 1 in $Y^{p,q}$. As in the proof of Proposition 4.3.3, we can find $t_1, \ldots, t_\ell \in T_{p,q}$, such that $x_i = t_1 t_2 \ldots t_i$ $(1 \le i \le \ell)$. Write $t_i = \Pi_q[\alpha_i]$ for a unique $\alpha_i \in S_p$ $(i = 1, \ldots, \ell)$. Then $[\alpha_1][\alpha_2] \ldots [\alpha_\ell]$ is a reduced word of length ℓ in Λ, since it is lifted from a path without backtracking; and because $\Pi_q([\alpha_1][\alpha_2] \ldots [\alpha_\ell]) = x_\ell = 1$, we see that $[\alpha_1][\alpha_2] \ldots [\alpha_\ell]$ belongs to $\Lambda(q)$. This proves that f_ℓ is the number of reduced words of length ℓ in Λ, belonging to $\Lambda(q)$.

Let $(x_0, x_1, x_2, x_3) \in \mathbb{Z}^4$ contribute to $s_Q(p^m)$, so $Q(x_0, x_1, x_2, x_3) = p^m$ and the correct congruences modulo 2 are satisfied. Form the quaternion $\alpha = x_0 + q(x_1 \, i + x_2 \, j + x_3 \, k)$: this α belongs to Λ' and, by Lemma 4.3.2, its equivalence class is in $\Lambda(q)$. From this we get the equality

$$s_Q(p^m) = |\{\alpha = a_0 + a_1 \, i + a_2 \, j + a_3 \, k \in \Lambda' : N(\alpha) = p^m, \, q \mid a_1, a_2, a_3\}|.$$

Suppose α contributes to the right-hand side of the previous equation. By Corollary 2.6.14, α has a unique factorization $\alpha = \pm p^\ell w_{m-2\ell}$, where $w_{m-2\ell}$ is a reduced word of length $m - 2\ell$ over S_p. The class $[\alpha]$ is therefore a reduced word of length $m - 2\ell$ in Λ that, moreover, belongs to $\Lambda(q)$. Conversely, starting from a reduced word w of length $m - 2\ell$ in $\Lambda(q)$, the formula $\alpha = \pm p^\ell w$ produces two quaternions as before. This shows that

$$|\{\alpha \in \Lambda' : N(\alpha) = p^m, \, [\alpha] \in \Lambda(q)\}| = 2 \sum_{0 \le r \le \frac{m}{2}} f_{m-2r},$$

which concludes the proof. $\square$

The trace formula for $X^{p,q}$ becomes, for every $m \in \mathbb{N}$:

$$s_Q(p^m) = \frac{2}{n} \, p^{\frac{m}{2}} \sum_{j=0}^{n-1} U_m\left(\frac{\mu_j}{2\sqrt{p}}\right).$$

At this juncture, we introduce the following subset Θ_p of $\mathbb{C}$:

$$\Theta_p = [i \log \sqrt{p}, 0] \cup [0, \pi] \cup [\pi, \pi + i \log \sqrt{p}].$$

Recall that the cosine and sine of a complex number $z \in \mathbb{C}$ are defined as

$$\cos z = 1 - \frac{z^2}{2!} + \frac{z^4}{4!} - \frac{z^6}{6!} + \cdots = \sum_{n=0}^{\infty} (-1)^n \frac{z^{2n}}{(2n)!} = \frac{e^{iz} + e^{-iz}}{2}$$

$$\sin z = z - \frac{z^3}{3!} + \frac{z^5}{5!} - \frac{z^7}{7!} + \cdots = \sum_{n=0}^{\infty} (-1)^n \frac{z^{2n+1}}{(2n+1)!} = \frac{e^{iz} - e^{-iz}}{2i}.$$

The reader can check easily that the change of variables $z \to 2\sqrt{p} \cos z$ maps Θ_p bijectively to $[-(p+1), p+1]$; note that it maps $[0, \pi]$ to $[-2\sqrt{p}, 2\sqrt{p}]$, so this change of variables "sees" the Ramanujan interval. For $j = 0, 1, \ldots,$ $n-1$, let $\theta_j \in \Theta_p$ be the unique element of Θ_p, such that $\mu_j = 2\sqrt{p} \cos \theta_j$. In particular, $\theta_0 = i \log \sqrt{p}$ and, if $\left(\frac{p}{q}\right) = -1$:

$$\theta_{n-1} = \pi + i \log \sqrt{p} \qquad \text{(by Corollary 4.3.6).}$$

By definition of the Chebyshev polynomial U_m, we have

$$s_Q(p^m) = \frac{2}{n} p^{\frac{m}{2}} \sum_{j=0}^{n-1} \frac{\sin(m+1)\theta_j}{\sin \theta_j}.$$

To prove that $X^{p,q}$ is Ramanujan, we, therefore, must prove that, with the exception of $\theta_0 = i \log \sqrt{p}$ and possibly of $\theta_{n-1} = \pi + i \log \sqrt{p}$, all the θ_j's are real. This was first done in [42], and we refer the reader to Remark 4.4.7 for an indication of how that proof was constructed. With elementary methods we will not be able to go so far. Instead, we will need to content ourselves with a proof that, for q sufficiently large, the imaginary part of θ_j is bounded above by a constant depending only on p. This will be enough to establish that the $X^{p,q}$ form a family of expanders.

Since, in the trace formula, the θ_j's are repeated according to the multiplicities of their corresponding eigenvalues, we first gather information about these multiplicities.

4.4.3. Proposition. Let μ be a nontrivial eigenvalue of $X^{p,q}$, which means that $|\mu| \neq p + 1$, and denote its multiplicity by $M(\mu)$. Then $M(\mu) \geq \frac{q-1}{2}$.

Proof. Let V_μ be the eigenspace corresponding to μ. From exercise 4 in section 4.1, the vector space V_μ is a representation space of the group underlying $X^{p,q}$. Since this group always contains $\mathrm{PSL}_2(q)$, V_μ is a representation space $\mathrm{PSL}_2(q)$. From Theorem 3.5.1, any nontrivial representation of $\mathrm{PSL}_2(q)$ has degree at least $\frac{q-1}{2}$. So we must prove that, if $|\mu| \neq p + 1$, then the representation of $\text{PSL}_2(q)$ on V_μ is nontrivial. We do this by contraposition, so we assume that the representation of $\text{PSL}_2(q)$ on V_μ is trivial. We separate two cases.

If $\left(\frac{p}{q}\right) = 1$, this means that every function in V_μ is constant, so $\mu = p + 1$.

If $\left(\frac{p}{q}\right) = -1$, then a nonzero function $f \in V_\mu$ must be constant on each of the two cosets of $\text{PSL}_2(q)$ in $\text{PGL}_2(q)$: say that

$$f = \begin{cases} a_+ & \text{on } \text{PSL}_2(q); \\ a_- & \text{on } \text{PGL}_2(q) - \text{PSL}_2(q). \end{cases}$$

Noting that f is an eigenfunction of the adjacency matrix of $X^{p,q}$, we are led to the system of equations:

$$\begin{cases} \mu a_- = (p+1)a_+ \\ \mu a_+ = (p+1)a_-. \end{cases}$$

Using the fact that f is nonzero, we get $\mu^2 = (p+1)^2$; hence, $|\mu| = p+1$, as desired. $\square$

We now reach the main result of this section.

4.4.4. Theorem. Fix a real number ε with $0 < \varepsilon < \frac{1}{6}$. For q sufficiently large, every nontrivial eigenvalue μ of $X^{p,q}$ satisfies

$$|\mu| \le p^{\frac{5}{6}+\varepsilon} + p^{\frac{1}{6}-\varepsilon}.$$

In particular, the $X^{p,q}$'s are a family of expanders.

Proof. We start with our expression for the trace formula for $X^{p,q}$:

$$s_\varrho(p^m) = \frac{2}{n} p^{\frac{m}{2}} \sum_{j=0}^{n-1} \frac{\sin(m+1)\theta_j}{\sin \theta_j},$$

for every $m \in \mathbb{N}$. Here $\mu_j = 2\sqrt{p} \cos \theta_j$. If μ_j is not in the Ramanujan interval $[-2\sqrt{p}, 2\sqrt{p}]$, we write

$$\begin{cases} \theta_j = i\psi_j & \text{if } 2\sqrt{p} < \mu_j \le p+1, \\ \theta_j = \pi + i\psi_j & \text{if } -(p+1) \le \mu_j < -2\sqrt{p}, \end{cases}$$

where $0 < \psi_j \le \log \sqrt{p}$ in both cases.

From now on, we assume that m is *even*. Recall that the hyperbolic sine and hyperbolic cosine of a complex number z are defined as

$$\sinh z = \frac{e^z - e^{-z}}{2} = i \sin(-iz);$$

$$\cosh z = \frac{e^z + e^{-z}}{2} = \cos(-iz).$$

For $\mu_j \notin [-2\sqrt{p}, 2\sqrt{p}]$, we have in both cases, since m is even,

$$\frac{\sin(m+1)\theta_j}{\sin \theta_j} = \frac{\sin i(m+1)\psi_j}{\sin i \psi_j} = \frac{\sinh(m+1)\psi_j}{\sinh \psi_j} \geq 0.$$

Then, for a fixed nontrivial eigenvalue $\mu_k \notin [-2\sqrt{p}, 2\sqrt{p}]$,

$$s_Q(p^m) = \frac{2}{n} p^{\frac{m}{2}} M(\mu_k) \frac{\sinh(m+1)\psi_k}{\sinh \psi_k} + \frac{2}{n} p^{\frac{m}{2}} \sum_{j:\mu_j \neq \mu_k} \frac{\sin(m+1)\theta_j}{\sin \theta_j}$$

$$\geq \frac{2}{n} p^{\frac{m}{2}} M(\mu_k) \frac{\sinh(m+1)\psi_k}{\sinh \psi_k} + \frac{2}{n} p^{\frac{m}{2}} \sum_{j:|\mu_j| \leq 2\sqrt{p}} \frac{\sin(m+1)\theta_j}{\sin \theta_j}.$$

We leave the reader to check that, for θ real, $\left| \frac{\sin(m+1)\theta}{\sin \theta} \right| \leq m+1$, so that

$$s_Q(p^m) \geq \frac{2}{n} p^{\frac{m}{2}} M(\mu_k) \frac{\sinh(m+1)\psi_k}{\sinh \psi_k} - 2 p^{\frac{m}{2}} (m+1).$$

We now estimate $s_Q(p^m)$ from before. By Remark 4.4.1, since m is even, $s_Q(p^m)$ is the number of integral solutions of

$$x_0^2 + 4q^2(x_1^2 + x_2^2 + x_3^2) = p^m.$$

We first estimate the number of possible choices for x_0. First, we have $|x_0| \leq p^{\frac{m}{2}}$. Second, $x_0^2 \equiv p^m \pmod{q^2}$; hence, by exercise 1 in Section 4.3,

$$x_0 \equiv \pm p^{\frac{m}{2}} \pmod{q^2}.$$

Since both x_0 and p are odd, we actually have

$$x_0 \equiv \pm p^{\frac{m}{2}} \pmod{2q^2}.$$

This gives at most two $\left(\frac{p^{\frac{m}{2}}}{q^2} + 1 \right)$ choices for x_0. Once x_0 is fixed, we must solve

$$x_1^2 + x_2^2 + x_3^2 = \frac{p^m - x_0^2}{4q^2}$$

in integers. Using the notations of section 2.2, there are $r_3\left(\frac{p^m-x_0^2}{4q^2}\right)$ ways to do this. Furthermore, by Corollary 2.2.13, we have

$$r_3\left(\frac{p^m-x_0^2}{4q^2}\right) = O_\varepsilon\left(\left(\frac{p^m}{q^2}\right)^{\frac{1}{2}+\varepsilon}\right)$$

for every $\varepsilon > 0$. Then,

$$s_Q(p^m) = O_\varepsilon\left[\frac{p^{\frac{m}{2}+\varepsilon m}}{q^{1+2\varepsilon}}\left(\frac{p^{\frac{m}{2}}}{q^2}+1\right)\right]$$

$$= O_\varepsilon\left[\frac{p^{m(1+\varepsilon)}}{q^{3+2\varepsilon}} + \frac{p^{\frac{m}{2}(1+2\varepsilon)}}{q^{1+2\varepsilon}}\right]$$

$$= O_\varepsilon\left[\frac{p^{m(1+\varepsilon)}}{q^3} + \frac{p^{\frac{m}{2}(1+2\varepsilon)}}{q}\right].$$

Thus, for some constant $C_\varepsilon > 0$, our inequality becomes

$$\frac{M(\mu_k)}{n}\cdot p^{\frac{m}{2}}\cdot\frac{\sinh(m+1)\,\psi_k}{\sinh\psi_k} \leq C_\varepsilon\left[\frac{p^{m(1+\varepsilon)}}{q^3} + \frac{p^{\frac{m}{2}(1+2\varepsilon)}}{q}\right] + p^{\frac{m}{2}}(m+1).$$

Canceling out $p^{\frac{m}{2}}$, and using $n \leq q^3$ (see Proposition 3.1.1), we get

$$M(\mu_k)\frac{\sinh(m+1)\,\psi_k}{\sinh\psi_k} \leq C_\varepsilon\left[p^{m(\frac{1}{2}+\varepsilon)}+q^2\,p^{m\varepsilon}\right]+q^3(m+1).$$

Suppose that m is chosen in such a way that $p^{\frac{m}{2}} \leq q^3$. Then

$$M(\mu_k)\frac{\sinh(m+1)\,\psi_k}{\sinh\psi_k} \leq C_\varepsilon\,[q^{3+6\varepsilon}+q^{2+6\varepsilon}]+q^3(1+6\log_p q).$$

Since $\sinh\psi_k \leq \sinh\log\sqrt{p}$, this yields

$$M(\mu_k)\sinh(m+1)\,\psi_k = O_\varepsilon\,[q^{3+6\varepsilon}].$$

Now take m to be the greatest even integer such that $p^{\frac{m}{2}} \leq q^3$. For q sufficiently large we have

$$\sinh(m+1)\,\psi_k \geq \frac{e^{(m+1)\psi_k}}{3} \geq \frac{e^{(-1+6\log_p q)\psi_k}}{3} \geq \frac{p^{-\frac{1}{2}}}{3}\,e^{6\log_p q\cdot\psi_k},$$

where we have used $\psi_k \leq \log\sqrt{p}$ in the final inequality. Then

$$M(\mu_k) = O_\varepsilon\left(q^{3+6\varepsilon-\frac{6\psi_k}{\log p}}\right).$$

But, since μ_k is a nontrivial eigenvalue, we have

$$M(\mu_k) \geq \frac{q-1}{2}$$

by Proposition 4.4.3. So, for q large enough, we must have

$$3 + 6\varepsilon - \frac{6\psi_k}{\log p} \geq 1, \text{ giving } \psi_k \leq \left(\frac{1}{3} + \varepsilon\right) \log p.$$

Then, since either $\theta_k = i \, \psi_k$ or $\theta_k = \pi + i \, \psi_k$, and $\mu_k = 2\sqrt{p} \cos \theta_k$, we get

$$|\mu_k| = 2\sqrt{p} \, |\cos(i \, \psi_k)| = 2\sqrt{p} \cosh \psi_k \leq p^{\frac{5}{6}+\varepsilon} + p^{\frac{1}{6}-\varepsilon},$$

for q big enough. This concludes the proof. $\square$

From Theorem 1.2.3 and Corollary 1.5.4, we immediately get estimates for the isoperimetric constant and the chromatic number of the graphs $X^{p,q}$.

4.4.5. Corollary. Fix $\varepsilon \in (0, 1/6)$. For q sufficiently large, one has

$$h(X^{p,q}) \geq \frac{p + 1 - p^{\frac{5}{6}+\varepsilon} - p^{\frac{1}{6}-\varepsilon}}{2}.$$

Moreover, if $\left(\frac{p}{q}\right) = 1$ and q is large enough,

$$\chi(X^{p,q}) \geq \frac{p+1}{p^{\frac{5}{6}+\varepsilon} + p^{\frac{1}{6}-\varepsilon}}.$$

This proves that we have given an explicit construction of an infinite family of graphs with large girth and large chromatic number, providing a constructive solution to the problem we discussed in section 1.6. There we used probabilistic methods to establish the existence of such graphs, but the proof gave no insight into how such a graph could be explicitly defined.

4.4.6. Corollary. Fix $N \in \mathbb{N}$. There exists an odd prime p, such that, for a prime q large enough,

$$g(X^{p,q}) \geq N \quad \text{and} \quad \chi(X^{p,q}) \geq N.$$

Proof. Let p be chosen large enough to have $\frac{p+1}{p^{\frac{11}{12}}+p^{\frac{1}{12}}} \geq N$. Then, choose q large enough to allow the following four conditions to be satisfied simultaneously:

(a) $q > p^8$;
(b) $2 \log_p q \geq N$;
(c) $\left(\frac{p}{q}\right) = 1$;
(d) $\chi(X^{p,q}) \geq \frac{p+1}{p^{\frac{11}{12}} + p^{\frac{1}{12}}}$.

(Condition (d) will be satisfied in view of Corollary 4.4.5.) Then, by Proposition 4.3.3 and Theorem 4.3.5, we have

$$\min\{g(X^{p,q}), \chi(X^{p,q})\} \geq N. \quad \square$$

4.4.7. Remark. The *Ramanujan conjecture* [54] is a conjecture about the order of magnitude of coefficients of modular cusp forms; in weight 2, this conjecture was proved by Eichler [23]. (For a discussion of modular forms, we refer the reader to [47].)

The θ-function of the quadratic form Q' is given by

$$\theta(z) = \sum_{x \in \mathbb{Z}^4} e^{2\pi i Q'(x)z} = \sum_{k=0}^{\infty} r_{Q'}(k) e^{2\pi i k z},$$

where $r_{Q'}(k)$ is the number of integral representations of the integer k by the form Q'. Then θ is a modular form of weight 2; decomposing θ as the sum of an Eisenstein series and a cusp form, we may appeal to Eichler's result to get estimates on $r_{Q'}(p^m) = s_Q(p^m)$ for even m's. Specifically, we get, for every $\varepsilon > 0$,

$$s_Q(p^m) = \frac{4}{q(q^2 - 1)} \frac{p^{m+1} - 1}{p - 1} + 0_\varepsilon \left(p^{\frac{m}{2}(1+\varepsilon)} \right).$$

(For details, see [42], [57], and [65].) It is interesting to compare this result to the estimates obtained in the proof of Theorem 4.4.4 by our elementary means.

Now, recall the trace formula for $X^{p,q}$, as written before Proposition 4.4.3:

$$s_Q(p^m) = \frac{2}{n} p^{\frac{m}{2}} \sum_{j=0}^{n-1} \frac{\sin(m+1)\theta_j}{\sin \theta_j}.$$

We leave as an easy exercise the proof that the dominant term $\frac{4}{q(q^2-1)} \frac{p^{m+1}-1}{p-1}$ is exactly the contribution of the trivial eigenvalues:

$$\begin{cases} \theta_0 = i \log \sqrt{p} & \text{if } \left(\frac{p}{q}\right) = 1; \\ \theta_0 = i \log \sqrt{p} \text{ and } \theta_{n-1} = \pi + i \log \sqrt{p} & \text{if } \left(\frac{p}{q}\right) = -1. \end{cases}$$

Suppose for simplicity that we are in the first case. Then, from the Ramanujan–
Eichler estimate on $s_Q(p^m)$, we get

$$\frac{2}{n} \sum_{j=1}^{n-1} \frac{\sin(m+1)\theta_j}{\sin\theta_j} = O_\varepsilon\left(p^{\frac{\varepsilon m}{2}}\right).$$

So, if some θ_j is not real, as in the proof of Theorem 4.4.4, we may write
$\theta_j = i\psi_j$ or $\theta_j = \pi + i\psi_j$, with $\psi_j \in (0, \log\sqrt{p}]$, and the corresponding
term is

$$\frac{2}{n} \frac{\sin(m+1)\theta_j}{\sin\theta_j} = \frac{2}{n} \frac{\sinh(m+1)\psi_j}{\sinh\psi_j} > 0,$$

since m is even. This quantity cannot cancel with the contributions of the real
θ_i's, since we certainly have

$$\left| \frac{2}{n} \sum_{i:\theta_i \text{ real}} \frac{\sin(m+1)\theta_i}{\sin\theta_i} \right| \le 2(m+1).$$

So, if some θ_j is not real, by choosing ε small enough, we get, for m a large
even number, a contradiction with the above estimate on $\sum_{j=1}^{n-1} \frac{\sin(m+1)\theta_j}{\sin\theta_j}$. This
way one proves that the graphs $X^{p,q}$ are Ramanujan.

Exercises on Section 4.4

1. Check that $z \mapsto 2\sqrt{p}\cos z$ maps Θ_p bijectively onto $[-(p+1), p+1]$.
2. Prove that, for θ real, $\left| \frac{\sin(m+1)\theta}{\sin\theta} \right| \le m+1$.
3. Show that $U_m\left(\frac{p+1}{2\sqrt{p}}\right) = p^{-\frac{m}{2}} \cdot \frac{p^{m+1}-1}{p-1}$.

4.5. Notes on Chapter 4

4.2 Theorem 4.2.2 is due to Lubotzky et al. [42], with a substantial part being obtained
independently by Margulis [46]. For applications of the graphs $X^{p,q}$ to problems
in automorphic forms, dynamical systems, and operator algebras, see [65].

4.3 The construction of the graphs $Y^{p,q}$ also appears in [42], [65], and [57]. When
$\left(\frac{p}{q}\right) = -1$, Biggs and Boshier [6] proved that the constant $c = \frac{4}{3}$ in
Corollary 4.3.6(b) is the best possible; namely

$$g(X^{p,q}) \le 4\log_p q + \log_p 4 + 2.$$

As far as the constant c in Definition 0.12 is concerned, examples nearly as good
as the $X^{p,q}$'s were constructed by Lazebnik, Ustimenko, and Woldar [38]; more
precisely, for every prime power q, they construct families $(X_m)_{m \in \mathbb{N}}$ of q-regular

graphs, such that

$$g(X_m) \geq \frac{4}{3} \, \log_q(q - 1) \cdot \log_{q-1} |X_m| \, .$$

4.4 For modular cusp forms of weight 2, the Ramanujan conjecture was proven by Eichler [23] as a consequence of Weil's proof of the Riemann hypothesis for curves over a finite field (see [68] and [69]).

He did so by relating the eigenvalues of certain operators acting on such spaces of cusp forms to the zeroes of zeta functions of modular curves over the fields $\mathbb{F}_p$. These operators are known as Hecke operators; they are close relatives of the operators A_r used in our setting in section 1.4. The Ramanujan property turns out to be equivalent to the Riemann Hypothesis for these zeta functions. In higher weight, the Ramanujan conjecture was proven by Deligne [18]. All these works rely heavily on algebraic geometry over finite fields.

Appendix
4-Regular Graphs with Large Girth

The aim of this Appendix is to give, following Margulis [45], a completely elementary construction of a family of 4-regular graphs with large girth, with an explicit estimate on the constant C in Definition 0.12. These graphs will be Cayley graphs of $SL_2(q)$, where q is an odd prime.

Let $\tau_q : SL_2(\mathbb{Z}) \to SL_2(q)$ denote reduction modulo q. In $SL_2(\mathbb{Z})$, consider the two matrices:

$$A = \begin{pmatrix} 1 & 2 \\ 0 & 1 \end{pmatrix}; \qquad B = \begin{pmatrix} 1 & 0 \\ 2 & 1 \end{pmatrix}.$$

Set then $A_q = \tau_q(A)$, $B_q = \tau_q(B)$, and define

$$S_q = \{A_q, A_q^{-1}, B_q, B_q^{-1}\},$$

so that S_q is a 4-element subset of $SL_2(q)$ (since q is odd). Set $X_q = \mathcal{G}(SL_2(q), S_q)$.

A.1.1. Lemma. For an odd prime q, the graph X_q is a 4-regular, connected graph on $q(q^2 - 1)$ vertices.

Proof. X_q is 4-regular, since $|S_q| = 4$; the number of vertices is given by Proposition 3.1.1(b). By Proposition 4.1.2(c), connectedness of X_q is equivalent to showing that S_q is a generating subset for $SL_2(q)$. To see the latter, we observe that, by Lemma 3.2.1, the matrices

$$\begin{pmatrix} 1 & 1 \\ 0 & 1 \end{pmatrix}, \begin{pmatrix} 1 & 0 \\ 1 & 1 \end{pmatrix}$$

generate $SL_2(q)$. Since $A_q^{\frac{q+1}{2}} = \begin{pmatrix} 1 & 1 \\ 0 & 1 \end{pmatrix}$ and $B_q^{\frac{q+1}{2}} = \begin{pmatrix} 1 & 0 \\ 1 & 1 \end{pmatrix}$, we see that S_q generates $SL_2(q)$. $\square$

To show that the graphs X_q are a family with large girth, we need some information about the subgroup H of $SL_2(\mathbb{Z})$ generated by A and B.

132

A.2.2. Proposition. H is isomorphic to the free group $\mathbb{L}_2$ on two generators.

Proof. H is the set of reduced words over the alphabet $\{A, A^{-1}, B, B^{-1}\}$; recall that a word is reduced if it contains no occurrence of AA^{-1}, $A^{-1}A$, BB^{-1}, $B^{-1}B$. Nonempty reduced words have exactly one form among the four following ones:

(a) words starting and finishing with a power of A:

$$A^{k_1} B^{\ell_1} A^{k_2} B^{\ell_2} \ldots A^{k_r} B^{\ell_r} A^{k_{r+1}},$$

where $k_i, \ell_i \in \mathbb{Z} - \{0\}$ ($1 \le i \le r+1, 1 \le j \le r$).

(b) words starting and finishing with a power of B:

$$B^{k_1} A^{\ell_1} B^{k_2} A^{\ell_2} \ldots B^{k_r} A^{\ell_r} B^{k_{r+1}},$$

where $k_i, \ell_j \in \mathbb{Z} - \{0\}$ ($1 \le i \le r+1, 1 \le j \le r$).

(c) words starting with a power of A and finishing with a power of B:

$$A^{k_1} B^{\ell_1} A^{k_2} B^{\ell_2} \ldots A^{k_r} B^{\ell_r},$$

where $k_i, \ell_j \in \mathbb{Z} - \{0\}$ ($1 \le i, j \le r$).

(d) words starting with a power of B and finishing with a power of A:

$$B^{k_1} A^{\ell_1} B^{k_2} A^{\ell_2} \ldots B^{k_r} A^{\ell_r},$$

where $k_i, \ell_j \in \mathbb{Z} - \{0\}$ ($1 \le i, j \le r$).

To prove that H is the free group over A and B, we have to show that any nonempty reduced word previously cited defines a nonidentity element in H. For this, we will use a method going back to Fricke and Klein [26], called today the "ping-pong lemma." Let $SL_2(\mathbb{Z})$ act on $\mathbb{R}^2$ by its standard linear action, and define two subsets E, F of $\mathbb{R}^2$ as follows (see Figure A.1):

$$E = \{(x, y) \in \mathbb{R}^2 : |y| > |x|\}$$

$$F = \{(x, y) \in \mathbb{R}^2 : |x| > |y|\}.$$

One sees immediately that $A^k(E) \subset F$ ($k \in \mathbb{Z} - \{0\}$) and $B^\ell(F) \subset E$ ($\ell \in \mathbb{Z} - \{0\}$). Now take a reduced word of the first kind from the previous:

$$W_1 = A^{k_1} B^{\ell_1} A^{k_2} B^{\ell_2} \ldots A^{k_r} B^{\ell_r} A^{k_{r+1}}.$$

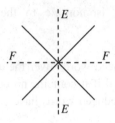

Figure A.1

To apply it to E, we start playing ping-pong:

$$A^{k_r+1}(E) \subset F$$
$$B^{\ell_r} A^{k_r+1}(E) \subset E$$
$$A^{k_r} B^{\ell_r} A^{k_r+1}(E) \subset F$$
$$B^{\ell_{r-1}} A^{k_r} B^{\ell_r} A^{k_r+1}(E) \subset E$$
$$\vdots$$

$$W_1(E) = A^{k_1} B^{\ell_1} \ldots A^{k_r} B^{\ell_r} A^{k_r+1}(E) \subset F.$$

Since $W_1(E) \subset F$ and $E \cap F = \emptyset$, clearly $W_1 \neq 1$. Proceeding symetrically, for a reduced word W_2 of the second form, we get $W_2(F) \subset E$ and therefore $W_2 \neq 1$.

Now let $W_3 = A^{k_1} B^{\ell_1} \ldots A^{k_r} B^{\ell_r}$ be a reduced word of the third kind. Choose $k \in \mathbb{Z}$ with $k \neq k_1$. Then $A^{-k} W_3 A_k$ is a word of the first kind; therefore, $A^{-k} W_3 A^k \neq 1$. Clearly this implies $W_3 \neq 1$. The case of a reduced word W_4 of the fourth kind is completely symmetric, so the proof is finished. $\square$

We endow $\mathbb{R}^2$ with the standard scalar product

$$\langle \xi \mid \eta \rangle = \xi_1 \eta_1 + \xi_2 \eta_2$$

(for vectors $\xi = (\xi_1, \xi_2)$, $\eta = (\eta_1, \eta_2)$ in $\mathbb{R}^2$); the corresponding euclidean norm is

$$\|\xi\| = \sqrt{\xi_1^2 + \xi_2^2}.$$

Let $M_2(\mathbb{R})$ be the space of 2-by-2 matrices with real coefficients. We define the *operator norm* of $T \in M_2(\mathbb{R})$ as

$$\|T\| = \sup \left\{ \frac{\|T \xi\|}{\|\xi\|} : \xi \in \mathbb{R}^2 - (0, 0) \right\}.$$

Then, by definition we have

$$\|T\,\xi\| \le \|T\|\,\|\xi\|$$

for every $\xi \in \mathbb{R}^2$. From this, it follows immediately that

$$\|T\,T'\,\xi\| \le \|T\|\,\|T'\|\,\|\xi\|$$

for $T, T' \in M_2(\mathbb{R})$; hence

$$\|T\,T'\| \le \|T\|\,\|T'\|.$$

A.3.3. Lemma. Let $T = \begin{pmatrix} a & b \\ c & d \end{pmatrix} \in M_2(\mathbb{R})$; denote by $T^t = \begin{pmatrix} a & c \\ b & d \end{pmatrix}$ the transposed matrix.

1. $\|T\| = \|T^t\|$;
2. $\|T\| = \|T^t\,T\|^{1/2}$;
3. $\|T\| \ge \max\{|a|, |b|, |c|, |d|\}$.
4. If T is symmetric ($b = c$) with eigenvalues $\lambda_1, \lambda_2 \in \mathbb{R}$, then $\|T\| = \max\{|\lambda_1|, |\lambda_2|\}$.

Proof. The proof is based on the following basic inequality, which follows immediately from the Cauchy–Schwarz inequality:

$$|\langle T\,\xi \mid \eta\rangle| \le \|T\|\,\|\xi\|\,\|\eta\|,$$

for every $\xi, \eta \in \mathbb{R}^2$.

1. Since $\langle T\,\xi \mid \eta\rangle = \langle \xi \mid T^t\,\eta\rangle$, we have

$$|\langle \xi \mid T^t\,\eta\rangle| \le \|T\|\,\|\xi\|\,\|\eta\|$$

for every $\xi, \eta \in \mathbb{R}^2$. Setting $\xi = T^t\,\eta$, we get

$$\|T^t\,\eta\|^2 \le \|T\|\,\|T^t\,\eta\|\,\|\eta\|,$$

from which we deduce

$$\|T^t\,\eta\| \le \|T\|\,\|\eta\|,$$

for every $\eta \in \mathbb{R}^2$; hence, $\|T^t\| \le \|T\|$. By symmetry we also have the reverse inequality.
2. We have $\|T^t\,T\| \le \|T^t\|\,\|T\| = \|T\|^2$, by (1). To prove the converse inequality, observe that, for every $\xi \in \mathbb{R}^2$,

$$\|T\,\xi\|^2 = \langle T\,\xi \mid T\,\xi\rangle = \langle T^t\,T\,\xi \mid \xi\rangle \le \|T^t\,T\| \cdot \|\xi\|^2,$$

by the basic inequality. So $\|T\|^2 \le \|T^t\,T\|$.

3. Set $\xi = (1, 0)$, $\eta = (0, 1)$. Then $a = \langle T \xi \mid \xi \rangle$, $b = \langle T \eta \mid \xi \rangle$, $c = \langle T \xi \mid \eta \rangle$, $d = \langle T \eta \mid \eta \rangle$. So the result follows immediately from the basic inequality.

4. If T is symmetric, it is conjugate to a diagonal matrix via an orthogonal matrix. Since conjugating by an orthogonal matrix does not change the operator norm, we may assume that T is diagonal; in other words, $T = \begin{pmatrix} \lambda_1 & 0 \\ 0 & \lambda_2 \end{pmatrix}$. Then, for $\xi = (\xi_1, \xi_2) \in \mathbb{R}^2 - \{(0, 0)\}$,

$$\frac{\|T \xi\|^2}{\|\xi\|^2} = \frac{\lambda_1^2 \xi_1^2 + \lambda_2^2 \xi_2^2}{\xi_1^2 + \xi_2^2} \leq \max\{\lambda_1^2, \lambda_2^2\},$$

so $\|T\| \leq \max\{|\lambda_1|, |\lambda_2|\}$. The converse inequality follows from (3).

$\square$

Example. As an application of the preceding lemma, let us compute the operator norm of $A = \begin{pmatrix} 1 & 2 \\ 0 & 1 \end{pmatrix}$, $B = \begin{pmatrix} 1 & 0 \\ 2 & 1 \end{pmatrix}$ and their inverses. We have

$$A^t A = \begin{pmatrix} 1 & 2 \\ 2 & 5 \end{pmatrix},$$

and the eigenvalues of $A^t A$ are $3 \pm 2\sqrt{2}$. Thus, we have $\|A^t A\| = 3 + 2\sqrt{2}$ and $\|A\| = \sqrt{3 + 2\sqrt{2}} = 1 + \sqrt{2}$, by Lemma A.3.3. Similarly,

$$\|A^{-1}\| = \|B\| = \|B^{-1}\| = 1 + \sqrt{2}.$$

A.4.4. Theorem. The graphs X_q, for q an odd prime, satisfy

$$\liminf_{q \to +\infty} \frac{g(X_q)}{\log_3 |X_q|} \geq \frac{1}{3 \log_3(1 + \sqrt{2})} = \frac{\log(3)}{3 \log(1 + \sqrt{2})} = 0.415492\ldots.$$

Proof. Write g for $g(X_p)$. By vertex-transitivity, X_q contains a circuit of length g starting and ending at $1 \in \mathrm{SL}_2(q)$:

$$x_0 = 1, x_1, \ldots, x_{g-1}, x_g = 1.$$

Since X_q is a Cayley graph, we find $\alpha_1, \alpha_2, \ldots, \alpha_g \in S_q$, such that $x_i = \alpha_1 \alpha_2 \ldots \alpha_{i-1} \alpha_i$ $(1 \leq i \leq g)$. Let $\widetilde{\alpha}_i$ be the unique element of $\{A, A^{-1}, B, B^{-1}\}$, such that $\tau_q(\widetilde{\alpha}_i) = \alpha_i$. Then $\widetilde{\alpha}_1 \widetilde{\alpha}_2 \ldots \widetilde{\alpha}_{g-1} \widetilde{\alpha}_g$ is a reduced word in H (since it is lifted from a circuit in X_q), and, since H is free (Proposition A.2.2), we have $\widetilde{\alpha}_1 \widetilde{\alpha}_2 \ldots \widetilde{\alpha}_{g-1} \widetilde{\alpha}_g \neq 1$ in H. On the other hand, since

$\tau_q(\tilde{\alpha}_1 \tilde{\alpha}_2 \ldots \tilde{\alpha}_g) = \alpha_1 \alpha_2 \ldots \alpha_g = x_g = 1$, we have

$$\tilde{\alpha}_1 \tilde{\alpha}_2 \ldots \tilde{\alpha}_g \in \operatorname{Ker} \tau_q.$$

This means that all coefficients of $\tilde{\alpha}_1 \tilde{\alpha}_2 \ldots \tilde{\alpha}_g - 1$ are divisible by q; since $\tilde{\alpha}_1 \tilde{\alpha}_2 \ldots \tilde{\alpha}_g - 1 \neq 0$, by Lemma A.3.3(3), we obtain

$$\|\tilde{\alpha}_1 \tilde{\alpha}_2 \ldots \tilde{\alpha}_g - 1\| \geq q.$$

By the triangle inequality: $\|\tilde{\alpha}_1 \tilde{\alpha}_2 \ldots \tilde{\alpha}_g\| \geq q - 1$. On the other hand, by the previous example, we also have

$$\|\tilde{\alpha}_1 \tilde{\alpha}_2 \ldots \tilde{\alpha}_g\| \leq (1 + \sqrt{2})^g.$$

Taking logarithms to the base 3, we get

$$\log_3(q - 1) \leq g \log_3(1 + \sqrt{2}).$$

By Lemma A.1,

$$\log_3(q - 1) = \frac{1}{3} \log_3 |X_q| + 0(1),$$

so that $g \geq \frac{1}{3 \log_3(1+\sqrt{2})} \log_3 |X_q| + 0(1)$, and the result follows. $\square$

The previous exposition owes much to an unpublished paper by P. de la Harpe, "Construction de 2 familles de graphes remarquables" (1989).

Exercises on Appendix

1. For $T \in M_2(\mathbb{R})$, why is $\|T\|$ finite? Check carefully that $T \to \|T\|$ is indeed a norm on $M_2(\mathbb{R})$.

2. Show that Theorem A.4.4 can be improved to

$$\liminf_{q \to +\infty} \frac{g(X_q)}{\log_3 |X_q|} \geq \frac{2}{3 \log_3(1 + \sqrt{2})},$$

by proceeding as follows: instead of working with $\tilde{\alpha}_1 \tilde{\alpha}_2 \ldots \tilde{\alpha}_g - 1$, work with $\tilde{\alpha}_1 \ldots \tilde{\alpha}_{[\frac{g}{2}]} - \tilde{\alpha}_g^{-1} \tilde{\alpha}_{g-1}^{-1} \ldots \tilde{\alpha}_{[\frac{g}{2}]+1}^{-1}$: all the coefficients of this matrix are divisible by q. Show that one has either $\|\tilde{\alpha}_1 \ldots \tilde{\alpha}_{[\frac{g}{2}]}\| \geq \frac{q}{2}$ or $\|\tilde{a}_g^{-1} \tilde{\alpha}_{g-1}^{-1} \ldots \tilde{\alpha}_{[\frac{g}{2}]+1}^{-1}\| \geq \frac{q}{2}$. Deduce that

$$\frac{q}{2} \leq (1 + \sqrt{2})^{\frac{g}{2}+1},$$

and conclude.

Bibliography

[1] M. AIGNER & G. M. ZIEGLER, Proofs from The Book, Springer-Verlag, Berlin/ New York, 1998.

[2] N. ALON, Eigenvalues and expanders, *Combinatorica* **6**(2) (1986), 83–96.

[3] N. ALON & V. MILMAN, λ_1, isoperimetric inequalities for graphs, and superconcentrators, *J. Comb. Theory*, Ser. B **38** (1985), 73–88.

[4] N. ALON & J. H. SPENCER, The probabilistic method, *Wiley-Interscience Series in Discrete Math and Optimization*, 1992.

[5] N. BIGGS, Algebraic graph theory, 2nd ed., Cambridge University Press, Cambridge, UK, 1994.

[6] N. L. BIGGS & A. G. BOSHIER, Note on the girth of Ramanujan graphs, *J. Comb. Theory*, Ser. B **49** (2) (1990), 190–194.

[7] B. BOLLOBAS, Extremal graph theory, *L.M.S. Monographs* **11**, Academic Press XX, 1978.

[8] R. BROOKS, Some relations between spectral geometry and number theory, *Topology* **90** (Colombus, OH, 1990), *Ohio State Univ. Math. Res. Inst. Publ.* **1**, de Gruyter (1992), 61–75.

[9] R. BROOKS & A. ZUK, On the asymptotic isoperimetic constants for Riemann surfaces and graphs, *Preprint*, Feb. 2001.

[10] M. BURGER, Cheng's inequality for graphs, *Preprint*, 1987.

[11] M. BURGER, Kazhdan constants for $SL_3(\mathbb{Z})$, *J. Reine Angew. Math.* **413**(1991), 36–67.

[12] P. BUSER, A note on the isoperimetric constant, *Ann. Sci. Ecole Norm. Sup.* **15** (1982), 213–230.

[13] J. CHEEGER, A lower bound for the smallest eigenvalue of the Laplacian, Probl. Analysis, Symposium in honor of S. Bochner, Princeton University (1970), 195–199.

[14] P. CHIU, Cubic Ramanujan graphs, *Combinatorica* **12**(1992), 275–285.

[15] F. R. K. CHUNG, Spectral graph theory, *Regional Conf. Ser. in Maths.* **92**, Amer. Math. Soc., 1997.

[16] Y. COLIN de VERDIÈRE, Spectres de graphes, *Cours spécialisés* **4**, Soc. Math. France, 1998.

[17] P. de la HARPE & A. VALETTE, La propriété (T) de Kazhdan pour les groupes localement compacts, *Astérisque* **175**, Soc. Math, France, 1989.

[18] P. DELIGNE, La conjecture de Weil I, *Publ. Math. IHES* **43**(1974), 273–308.

[19] L. E. DICKSON, Arithmetic of quaternions, *Proc. London Math. Soc.* **20**(2) (1922), 225–232.

[20] L. E. DICKSON, Linear groups with an exposition of the Galois field theory, Dover Publications, New York, 1958.

[21] P. G. L. DIRICHLET, Sur l'équation $t^2 + u^2 + v^2 + w^2 = 4m$, *J. Math. pures et appliquées* **1**(1856), 210–214.

[22] J. DODZIUK, Difference equations, isoperimetric inequality and transience of certain random walks, *Trans. Amer. Math. Soc.* **284**(1984), 787–794.

[23] M. EICHLER, Quaternäre quadratische Formen und die Riemannsche Vermutung für die Kongruenz zeta funktion, *Arch. Math.* **5**(1954), 355–366.

[24] P. ERDÖS, Graph theory and probability, *Can. J. Math.* **11**(1959), 34–38.

[25] P. ERDÖS & H. SACHS, Reguläre Graphen gegebener Taillenweite mit minimaler Knollenzahh, *Wiss. Z. Univ. Halle-Willenberg Math. Nat. R.* **12**(1963), 251–258.

[26] R. FRICKE & F. KLEIN, Vorlesungen über die Theorie der elliptischen Modulfunctionen, Teubner, Leipzig, 1890.

[27] G. FROBENIUS, Über Gruppencharaktere, Sitzungsberichte der Königlich Preußischen Akademie der Wissenschaften zu Berlin (1896), 985–1021.

[28] O. GABBER & Z. GALIL, Explicit constructions of linear-sized superconcentrators, *J. Comput. Syst. Sci.* **22**(1981), 407–420.

[29] A. GAMBURD, On spectral gap for infinite index "congruence" subgroups of $SL_2(\mathbb{Z})$, *Preprint.*

[30] O. GOLDREICH, R. IMPAGLIAZZO, L. LEVIN, R. VENKATESAN & D. ZUCKERMAN, Security preserving amplification of hardness, *31st annual symp. on foundations of computer science*, Vol. 1 (1990), 318–326.

[31] R. I. GRIGORCHUK & A. ZUK, On the asymptotic spectrum of random walks on infinite families of graphs, Eds. M. Picardello et al., Random walks and discrete potential theory, Cortona 1997. Cambridge University Press. *Symp. Math.* **39**(1999), 188–204.

[32] G. H. HARDY & E. M. WRIGHT, An introduction to the theory of numbers, 5th ed., Clarendon Press, Oxford, 1979.

[33] A. J. HOFFMAN, On eigenvalues and colorings of graphs, in *Graph Theory and its Applications*, ed. B. Harris, Academic Press, New York, 1970, pp. 79–91.

[34] B. HUPPERT, Endliche Gruppen I, *Grundlehren der Math. Wiss.* **134**, Springer-Verlag, Berlin/New York, 1979.

[35] A. HURWITZ, Vorlesungen über die Zahlentheorie der Quaternionen, *J. Springer*, Berlin, 1919.

[36] H. KESTEN, Symmetric random walks on groups, *Trans. Amer. Math. Soc.* **92**(1959), 336–354.

[37] E. LANDAU, Elementary Number Theory, 2nd ed., translation by J. Goodman, Chelsea Publishing Co, New York, 1966.

[38] F. LAZEBNIK, V. A. USTIMENKO & A. J. WOLDAR, A new series of dense graphs of high girth, *Bull. Amer. Math. Soc., New Ser.* **32**(1) (1995), 228–239.

[39] W. C. W. LI, Number theory with applications, *Series on Univ. Math.* **7**, World Scientific (Singapore), 1995.

[40] W. C. W. LI & P. SOLÉ, Spectra of regular graphs and hypergraphs and orthogonal polynomials, *Eur. J. Comb.* **17**(5) (1996), 461–477.

[41] A. LUBOTZKY, Discrete groups, expanding graphs and invariant measures, *Progress in Mathematics* **125**, Birkhaeuser, Basel, 1994.

[42] A. LUBOTZKY, R. PHILLIPS, & P. SARNAK, Ramanujan graphs, *Combinatorica* **8** (1988), 261–277.

[43] A. V. MALYSĚV, Ueber die Darstellung ganzer Zahler mittels positiver quadratischer Formen, *Tr. Math. Inst. Steklov* **65**(1962), 1–212.

[44] G. A. MARGULIS, Explicit construction of concentrators, *Problems Inform. Transmission* **9**(1973), 325–332.

[45] G. A. MARGULIS, Explicit constructions of graphs without short cycles and low density codes, *Combinatorica* **2**(1982), 71–78.

[46] G. A. MARGULIS, Explicit group-theoretical constructions of combinatorial schemes and their application to the design of expanders and concentrators, *J. Probl. Inf. Transm.* **24**(1) (1988), 39–46.

[47] T. MIYAKE, Modular forms, Springer-Verlag, Berlin/New York, 1989.

[48] M. MORGENSTERN, Existence and explicit constructions of $q + 1$ regular Ramanujan graphs for every prime power q, *J. Comb. Theory*, Ser. B **62** (1994), 44–62.

[49] M. A. NAIMARK & A. I. ŠTERN, Theory of group representations, *Grundlehren der Math. Wiss.* **246**, Springer-Verlag, Berlin/New York, 1982.

[50] J. NAOR & M. NAOR, Small-bias probability spaces: efficient constructions and applications, *SIAM J. Comput.* **22**(1993), 838–856.

[51] A. NILLI, On the second eigenvalue of a graph, *Discrete Math.* **91**(1991), 207–210.

[52] M. S. PINSKER, On the complexity of a concentrator, in *7th International Teletraffic Conference*, Stockholm, June 1973, 318/1-318/4.

[53] N. PIPPENGER, Sorting and selecting in rounds, *SIAM J. Comput.* **16**(1987), 1032–1038.

[54] S. RAMANUJAN, On certain arithmetical functions, *Trans. Cambridge Phil. Soc.* XXII **9**(1916), 159–184.

[55] O. REINGOLD, S. VADHAN, & A. WIGDERSON, Entropy waves, the zig-zag graph product and constant-degree expanders, *Annals of Math.*, Jan. 2002.

[56] P. SAMUEL, Théorie algébrique des nombres, 2ème éd., Coll. Méthodes, Hermann, Paris, 1971.

[57] P. SARNAK, Some applications of modular forms, *Cambridge Trusts in Math.* **99**, Cambridge University Press, Cambridge, UK, 1990.

[58] P. SARNAK, Selberg's eigenvalue conjecture, *Notices Amer. Math. Soc.* **42**(11) (1995), 1272–1277.

[59] P. SARNAK & X. XUE, Bounds for multiplicities of automorphic representations, *Duke Math. J.* **64**(1) (1991), 207–227.

[60] J.-P. SERRE, Représentations linéaires des groupes finis, 5ème éd., Collection Méthodes, Hermann, Paris, 1998.

[61] J.-P. SERRE, Cours d'arithmétique, 2ème éd., Coll. Le Mathématicien, Presses Universitaires de France, 1977.

[62] J.-P. SERRE, Répartition asymptotique des valeurs propres de l'opérateur de Hecke T_p, *J. Am. Math. Soc.* **10**(1) (1997), 75–102.

[63] M. SIPSER & D. A. SPIELMAN, Expander codes, *IEEE Trans. Inf. Theory* **42**(1996), 1710–1722.

[64] M. SUZUKI, Group theory, *Grundlehren der Math. Wiss.* **247**, Springer-Verlag, Berlin/New York, 1982.

[65] A. VALETTE, Graphes de Ramanujan et applications, Séminaire Bourbaki, Volume 1996/97, Exposés 820–834, *Astérisque* **245**, Soc. Math. France, 1997.

[66] L. G. VALIANT, Graph-theoretic arguments in low-level complexity, *Math. Found. Comput. Sci., Proc. 6th Symp.*, Tatranska Lomnica 1977, *Lect. Notes Comput. Sci.* **53**(1977), 162–176.

[67] A. WEIL, Sur les sommes de trois et quatre carrés (1974), Œuvres Scientifiques Vol. III, Springer-Verlag, Berlin/New York, 1979.

[68] A. WEIL, Sur les courbes algébriques et les variétés qui s'en déduisent, Hermann, Paris, 1948.

[69] A. WEIL, On some exponential sums, *Proc. Nat. Acad. Sci. USA* **34**(1948), 204–207.

[70] A. WIGDERSON & D. ZUCKERMAN, Expanders that beat the eigenvalue bound: explicit construction and applications, *Combinatorica* **19**(1999), 125–138.

Index